WASTE AND WASTE MANAGEMENT

RADIOACTIVE WASTES AND EXPOSURE

FUNDAMENTALS, MANAGEMENT STRATEGIES AND ENVIRONMENTAL IMPLICATIONS

Waste and Waste Management

Additional books in this series can be found on Nova's website under the Series tab.

Additional e-books in this series can be found on Nova's website under the eBooks tab.

WASTE AND WASTE MANAGEMENT

RADIOACTIVE WASTES AND EXPOSURE

FUNDAMENTALS, MANAGEMENT STRATEGIES AND ENVIRONMENTAL IMPLICATIONS

AUSTIN D. RUSSELL
EDITOR

Copyright © 2017 by Nova Science Publishers, Inc.

All rights reserved. No part of this book may be reproduced, stored in a retrieval system or transmitted in any form or by any means: electronic, electrostatic, magnetic, tape, mechanical photocopying, recording or otherwise without the written permission of the Publisher.

We have partnered with Copyright Clearance Center to make it easy for you to obtain permissions to reuse content from this publication. Simply navigate to this publication's page on Nova's website and locate the "Get Permission" button below the title description. This button is linked directly to the title's permission page on copyright.com. Alternatively, you can visit copyright.com and search by title, ISBN, or ISSN.

For further questions about using the service on copyright.com, please contact:
Copyright Clearance Center
Phone: +1-(978) 750-8400   Fax: +1-(978) 750-4470   E-mail: info@copyright.com.

NOTICE TO THE READER

The Publisher has taken reasonable care in the preparation of this book, but makes no expressed or implied warranty of any kind and assumes no responsibility for any errors or omissions. No liability is assumed for incidental or consequential damages in connection with or arising out of information contained in this book. The Publisher shall not be liable for any special, consequential, or exemplary damages resulting, in whole or in part, from the readers' use of, or reliance upon, this material. Any parts of this book based on government reports are so indicated and copyright is claimed for those parts to the extent applicable to compilations of such works.

Independent verification should be sought for any data, advice or recommendations contained in this book. In addition, no responsibility is assumed by the publisher for any injury and/or damage to persons or property arising from any methods, products, instructions, ideas or otherwise contained in this publication.

This publication is designed to provide accurate and authoritative information with regard to the subject matter covered herein. It is sold with the clear understanding that the Publisher is not engaged in rendering legal or any other professional services. If legal or any other expert assistance is required, the services of a competent person should be sought. FROM A DECLARATION OF PARTICIPANTS JOINTLY ADOPTED BY A COMMITTEE OF THE AMERICAN BAR ASSOCIATION AND A COMMITTEE OF PUBLISHERS.

Additional color graphics may be available in the e-book version of this book.

Library of Congress Cataloging-in-Publication Data

ISBN: 978-1-53612-213-8

Published by Nova Science Publishers, Inc. † New York

CONTENTS

Preface   **vii**

Chapter 1   Radiation Exposure in Medical Imaging: Risks, Measurements and Health Effects   **1**
Danijela Arandjic, Predrag Bozovic, Olivera Ciraj-Bjelac, Sandra Ceklic, Djordje Lazarevic and Jelena Stankovic Petrovic

Chapter 2   Solid Radioactive Wastes in Nuclear Medicine   **65**
Gabriela Hoff and Cláudia R. Brambilla

Chapter 3   The Calculation of Dosimetry in Small Animals Combining the FDTD Method and Experimental Measures: Application for Three Radiation Systems   **127**
Elena López-Martín, Aarón A. Salas-Sánchez, Alberto López-Furelos, Francisco J. Jorge-Barreiro, Eduardo Moreno-Piquero and Francisco J. Ares-Pena

Chapter 4	Evolution of the International Law on Wastes Disposal in Geological Environments in BRICS *Daniel Figueira de Barros and* *Daniel Marcos Bonotto*	**147**
Chapter 5	Assessments of Local Acceptance of Radioactive Waste Facilities *Taehyun Kim*	**179**
Index		**201**

PREFACE

Medical exposure to ionizing radiation remains by far the largest contributor to the radiation dose from manmade sources of radiation. Strategies for dose and risk management, with a focus on the minimization of the stochastic effect and prevention of tissue reactions are presented in Chapter One. A critical review about solid radioactive wastes in Nuclear Medicine facilities and its management, storage and disposal is covered in Chapter Two. Chapter Three presents an assessment of a mixed system for calculating electromagnetic dosimetry that combined experimental radiation parameters and Finite Differences in Temporal Domain simulations based on numerical phantom rats. Chapter Four described and compares the legislation and legal provisions on waste disposal for medium, low and high levels of radiation among BRICS countries – Brazil, Russia Federation, India, People's Republic of China and South Africa. Chapter Five introduces studies on the assessment of local acceptance of radioactive waste facilities.

Chapter 1 - Medical exposure to ionizing radiation remains by far the largest contributor to the radiation dose from manmade sources of radiation. Current issues in radiation protection in medical exposure include not only the rapidly increasing collective dose, but also that a large number of diagnostic imaging procedures are unnecessary. The use of relatively higher dose examinations and procedures necessitates the increased emphasis of application of fundamental radiation protection principles as a form of justification and optimization. Diagnostic and interventional procedures cover a diverse range of examination types, many of which are increasing in frequency and technical complexity. The demand for radiation dose and risk assessment has resulted in the development of new dosimetric measuring instruments, techniques and terminologies which affect work in the clinical environment and calibration facilities. Objectives of clinical dose measurements in diagnostic and interventional radiology are described, as well as requirements for dosimeters and procedures to assess dose to standard dosimetry phantoms and patients in clinical diverse modalities. High dose modalities such as computed tomography, interventional procedures and sensitive population groups such as children are of particular concern. Strategies for dose and risk management, with a focus on the minimization of the stochastic effect and prevention of tissue reactions are presented.

Chapter 2 - This is a critical review about solid radioactive wastes in Nuclear Medicine facilities and its management, storage and disposal. In the last 10 years, the number of nuclear medicine facilities is growing up and the radioactive wastes need to be managed in a correct way to prevent environmental implications in their disposal. Storage of radioactive waste in medical facilities is a procedure to deal with relatively short physical half-life radionuclides (such as 8 days, 6 hours or 110 minutes to Iodine 131, Technetium 99m and Fluorine 18 respectively). This approach depends on many factors, such as segregation, type of radionuclide, mass, dose rate and residual activity, time of storage and date of disposal. Radioactive solid wastes are present in different kinds of medical materials in a Nuclear Medicine facility and are generally classified as low-level radioactive waste. Many countries have defined previously clearance levels for radionuclides and some of them are referred to statistically significant differences from background activity. It is the responsibility of the regulatory authority of each country to define clearance levels and site-specific discharge authorizations. Different directives around the world contain a list of nuclides with values of quantities (Bq) and concentrations of activity per unit mass (Bq/g) that should not to be exceeded in radioactive waste disposal. The practice in some countries for this procedure of estimating this quantity/activity may be based on theoretical calculations and estimating by using Geiger Müller measurement of exposure rate or the total storage time is based in 10% of the initial activity. This measure estimation using GM could be affected by many factors as energy dependence of the detector, radionuclide involved - energy spectra, geometry of the solid waste and it heterogeneity of the waste container or box, mixture of radionuclides in the segregation process, among others factors. In this critical review is presented the international regulations and procedures for management of solid radioactive waste generated in Nuclear Medicine and its evolution in time according as well as the current practice in different countries. It will presented results of Monte Carlo simulations and deterministic calculation to perform an evaluation of different factors that could affect the determination of activity concentration in the management of solid radioactivity wastes and its implications to discharge it correctly based on geometry of the solid radioactive waste container/box when is used Geiger Müller detectors.

Chapter 3 - This chapter presents an assessment of a mixed system for calculating electromagnetic dosimetry that combined experimental radiation parameters and Finite Differences in Temporal Domain (FDTD) simulations based on numerical phantom rats. Using a simple formula that accounts for the different radiation exposure variables, it was possible to determine the specific absorption rate (SAR) in tissues of animals exposed to non-ionizing radiation in three experimental systems (standing wave cavity, traveling wave cavity and multifrequency). To estimate the SAR values for these three experimental systems, it was necessary to combine the measured values of the power absorbed by the animal (in the standing wave cavity) or the value of the |E| field (a Gigahertz Transverse Electromagnetic (GTEM) chamber was used for traveling wave, single frequency, and multifrequency experiments) with FDTD numerical computations. The three experimental systems were analyzed and different biological models were compared. The authors also establish a discussion on the methodological evolution of dosimetry calculation and the biological results obtained in these three experimental radiation systems.

Chapter 4 - The world's concern over where to store radioactive waste gained prominence in 1978 at a technical committee meeting held in London coordinated by the International Atomic Energy Agency of the United Nations (IAEA), in order collect information about regulations and experiences from several attendee countries as well as they addressed regulatory approaches, thus there was a large discussion on the various aspects and issues involved in the subject. Primordially, at the time it was defined that repository systems were to include the burial of wastes in either shallow or deep depths, also the disposition of radioactive wastes should take place in caves and in continental geological formations. At present, the final disposal of radioactive wastes in a safe way, in the light of the current, scientific and technological development, allows two main possible final destinations: the disposal in the environment or the confinement into the so-called final repositories. Confinement implies the definitive waste isolation inside a repository for long periods of time – from hundreds to thousands of years, depending on the half-life of the radionuclide. The purpose of this chapter is to describe and compare the legislation and legal provisions on waste disposal for medium, low and high levels of radiation among BRICS countries – Brazil, Russia Federation, India, People's Republic of China and South Africa. This chapter compares regulatory provisions, legislation, international agreements and competences of entities and/or international regulatory governmental bodies, showing BRIC'S status of law considering if their laws need improvement and reconsideration. Moreover, the chapter aims to report, in a short way, the historical evolution of international law on waste disposal in geological environment in BRICS by means of research on the international documentation and analysis based on bulletins from the Nuclear Energy Agency (NEA), a multinational and intergovernmental body part of the Organization for Economic Co-operation and Development (OECD) affiliated to the IAEA.

Chapter 5 - The purpose of this chapter is to introduce studies on the assessment of local acceptance of radioactive waste facilities (RWFs) using different methods. Because radioactive waste disposal facilities (RWDFs) are one of the most controversial locally unwanted land uses, siting these facilities near human habitation has been a growing issue in urban planning and environmental management. Some studies on the siting issue have been conducted using engineering and geographical approaches. Others considered social and economic perspectives using quantitative and qualitative methods. However, there is a lack of literature combining these different approaches. Four studies introduced in this chapter measured local acceptance for low- and high-level RWDFs using qualitative, quantitative, and mixed methods approaches.

The first two studies analyzed the spatial patterns of the referendum results for siting a low-level RWDF. The facility was assigned to be placed in Gyeongju city after a competitive local referendum amongst four candidate cities in Korea in 2005. However, many conflicts between the residents living within and near Gyeongju occurred after the decision. By analyzing spatial patterns of the referendum data, the first study identified that the local acceptance near and far from the facility were clustered with different values. The results of face-to-face interviews showed that people near the nuclear power plant had low risk perception, and the benefit of financial compensation for the districts offset the cost of potential risk. The second study analyzed the spatial distribution of the acceptance rate in each ward of the four candidate cities using several spatial statistical methods and several types of interviews. The results showed that the referendum system had a problem with spatial inequity within and across its jurisdiction.

The last two studies examined the local acceptance for a high-level RWDF using qualitative and quantitative methods. As the temporary storage for spent nuclear fuel in Korea has almost reached its limit, the question of relocating the facility has become urgent. Because perceptions on high-level radioactive waste may be higher compared to those on low-level radioactive waste, the latter two studies identified perception types and factors of local acceptance for spent nuclear fuel repository. One study identified four types of local acceptance—safety concerns–government distrust, safety trust–government trust, safety concerns–conflict avoidance, and citizen participation—and differences among these perception types using Q methodology. The other study conducted a survey and identified five factors of residents' perception of spent nuclear fuel repository by analyzing the structural equation model. The results showed that environmental impacts and economic feasibility had a high positive relation to local acceptance rather than risk perception. In particular, environmental impacts were distinctively high regardless of demographic characteristics.

These examples may provide new perspectives on management strategies of locating RWFs regarding local acceptance of these facilities.

In: Radioactive Wastes and Exposure
Editor: Austin D. Russell

ISBN: 978-1-53612-213-8
© 2017 Nova Science Publishers, Inc.

Chapter 1

RADIATION EXPOSURE IN MEDICAL IMAGING: RISKS, MEASUREMENTS AND HEALTH EFFECTS

Danijela Arandjic, Predrag Bozovic, Olivera Ciraj-Bjelac[*], Sandra Ceklic, Djordje Lazarevic and Jelena Stankovic Petrovic

Radiation and Environmental Protection Department
Vinca Institute of Nuclear Sciences, University of Belgrade
Belgrade, Serbia

ABSTRACT

Medical exposure to ionizing radiation remains by far the largest contributor to the radiation dose from manmade sources of radiation. Current issues in radiation protection in medical exposure include not only the rapidly increasing collective dose, but also that a large number of diagnostic imaging procedures are unnecessary. The use of relatively higher dose examinations and procedures necessitates the increased

---

[*] Corresponding Author Email: ociraj@vinca.rs.

emphasis of application of fundamental radiation protection principles as a form of justification and optimization. Diagnostic and interventional procedures cover a diverse range of examination types, many of which are increasing in frequency and technical complexity. The demand for radiation dose and risk assessment has resulted in the development of new dosimetric measuring instruments, techniques and terminologies which affect work in the clinical environment and calibration facilities. Objectives of clinical dose measurements in diagnostic and interventional radiology are described, as well as requirements for dosimeters and procedures to assess dose to standard dosimetry phantoms and patients in clinical diverse modalities. High dose modalities such as computed tomography, interventional procedures and sensitive population groups such as children are of particular concern. Strategies for dose and risk management, with a focus on the minimization of the stochastic effect and prevention of tissue reactions are presented.

Keywords: radiation dose, medical exposure, radiation protection, risk assessment

INTRODUCTION

Although the terms "radiation," "radioactive" or "nuclear" cause fear and concern, the radiation is a part of our natural environment and contributes greatly to the quality of life and to the development of science, industry and medicine. Meanwhile, radiation saves lives; medical exposure to ionizing radiation remains by far the largest contributor to the radiation dose from manmade sources of radiation.

The use of x-ray imaging in medicine continues to grow, bringing significant benefits for the population in both diagnosis and management of disease [1, 2]. Millions of x-ray examinations, both diagnostic and interventional, nuclear medicine procedures and radiotherapy treatments are performed annually worldwide. According to the current analysis, approximately 3.6 billion diagnostic x-ray examinations (including diagnostic medical and dental examinations), 33 million nuclear medicine examinations and more than 5 million radiotherapy treatments are undertaken annually throughout the world [1].

Inevitably, medical exposure has grown very rapidly over the last three decades in many countries. It is important to note that the use of radiation in medicine has significantly increased in the past decade in terms of the number of examinations, collective dose and per caput dose [1, 3]. During last couple of decades, in some low-dose imaging modalities, as radiography, progress in technology has resulted in dose reduction. However, this has not been the case with high-dose imaging modalities such as computed tomography (CT) and interventional procedures [1, 4, 5, 6, 7]. The main reason for such growth lays in the fact that modern medicine demands rapid diagnosis and treatment. Radiology and other modalities based on ionizing radiation are an essential component of patient management. In addition, many patients have multiple investigations using ionizing radiation, from which particularly CT or fluoroscopy guided procedures require special attention.

Diagnostic and interventional radiology has been undergoing rapid changes in technology and practices that have implications on radiation safety of the patients [8-11]. In particular, after the rapid adoption of multi-detector CT, radiation doses from CT are now the single largest source of diagnostic radiation exposure to patients [1, 4]. In the past two decades, the use of CT scanning has increased by more than 800% globally. In particular, in the United States, over the period of 1991 to 2002, a 10–20% growth per year in CT use [4]. The National Council on Radiation Protection and Measurements reported the changes of ionizing radiation exposure of the population of the United States. The expected average radiation exposure in the US was 3.6 mSv in the 1980s, but increased to 6.2 mSv in 2006, making medical exposure comparable with the exposure due to natural background radiation. In particular, CT was the most important contributor, at 49% of all medical exposure. Nuclear medicine contributes 26% of all medical exposure, whereas exposure resulting from radiation therapy was not included in the analysis [2].

In addition to development of x-ray modalities, the range of radionuclides that can be used in nuclear medicine has also increased. Despite huge benefits, some of those procedures deliver high radiation doses to patients, both in diagnostic and interventional radiology and nuclear medicine and hybrid imaging. Increased use of radiation in medicine resulted in occurrence of radiation injuries in interventional radiology and cardiology, and accidental exposures in radiotherapy. Fortunately, such situations are not common compared to the number of procedures or treatments performed, but were increasingly reported in the 1990s and 2000s [5].

RADIATION PROTECTION PRINCIPLES APPLIED TO MEDICAL EXPOSURE

International Commission on Radiological Protection (ICRP) proposed a system of radiation protection with its three principles of justification, optimization and individual dose limitation in the Publication 26 [12]. In the Publication 60, ICRP revised its recommendations and extended its philosophy to a system of radiological protection while keeping the fundamental principles of protection [13]. ICRP published report 103, as a revised general recommendation for a system of radiation protection in 2007 [14, 15]. This new recommendation provides guidance on the fundamental principles on which appropriate radiological protection in different exposure situation, including medical exposures, can be based. To underpin these ICRP recommendations with regard to the medical exposure of patients, the ICRP has published the Publication 105, [16], in which detailed advice related to radiological protection and safety in the medical applications of ionizing radiation has been consolidated.

At present, ICRP suggested three general principles of radiation protection as: justification, optimization and dose limit. Given unique considerations related to medical exposure, e.g., exposure of patients for diagnostic and therapeutic purposes, it is not appropriate to apply dose limits or dose constraints. However, to manage and control radiation exposure risks, any medical radiation exposure must be justified. Once justified, the examinations which use ionizing radiation must be optimized. Justification means that the examination must be apocopate. Optimization means that the imaging should be performed using doses that are as low as reasonably achievable, consistent with the diagnostic task.

Current issues in radiation protection in medical exposure include not only the rapidly increasing collective dose, but also that a large number of diagnostic imaging procedures are unnecessary. The use of relatively higher dose examinations and procedures is going to necessitate increased emphasis of application of above mentioned radiation protection principles as justification and optimization.

Justification of Radiological Practice in Medicine

The decision to adopt or continue any human activity involves a review of the benefits and disadvantages of the possible options. This review usually provides a number of alternative procedures that will do more good than harm [16]. The need for better justification emerged from substantial increase in the contribution of medical imaging to collective dose, as described above [2].

The benefits of practices and procedures that utilize ionizing radiation are well established and well accepted both by the health professionals and society as a whole. When a procedure involving radiation is medically justified, the anticipated benefits are almost always identifiable and well understood. On the other hand, the risk of adverse effects is often difficult to estimate. In its Publication 103, ICRP stated as a principle of justification that "Any decision that alters the radiation exposure situation should do more good than harm" [14, 16].

The ICRP has recommended a multi-step approach to the justification of patient exposures that identifies three levels at which justification operate [14, 16, 17].

Level 1 deals with the use of radiation in medicine in general. In practice, such use is accepted as doing more good than harm to the patient, and its justification is taken for granted. Level 2 deals with specified procedures with a specified objective (e.g., chest radiographs for patients showing relevant symptoms). The aim at this level is to judge whether the procedure will improve diagnosis or provide necessary information about those exposed. Finally, Level 3 deals with the application of the procedure to an individual (i.e., whether the particular application is judged to do more good than harm to the individual patient). In practice, all individual medical exposures need to be justified in advance, by taking into account the specific objectives of the exposure and the characteristics of the individual patient [16, 17].

In the case of the individual patient, justification normally involves both the referring medical doctor (who refers the patient, as patient's physician and surgeon familiar with the patient and the medical history) and radiologist or nuclear medicine specialist under whose responsibility the examination is conducted. In general, the later takes overall responsibility for the examination and needs to work in close cooperation with the referring physician(s) in order to establish the most appropriate procedure for the management of the patient. This is of particular importance for some specific population groups, as children and female patients of reproductive capacity. Although important for all patient categories, it is particularly important with infants and children that the feasibility of alternative techniques that do not use ionizing radiation, as ultrasound and magnetic resonance imaging (MRI) is considered.

A number of tools are available to effectively implement justification principle trough selection of correct radiological examination for a particular patient. The most widely known involve "appropriateness or referral criteria and/or guidelines." Referral guidelines, often proposed by professional societies, provide advice on the appropriateness of imaging modalities and specific examinations for many common clinical presentations and help to exclude inappropriate examinations [17-19]. Imaging referral guidelines have been available for over 20 years in Europe. The Royal College of Radiologists (RCR) first published a guideline: "Making the Best Use of a Department of Clinical Radiology" in 1989 [20]. The Radiation Protection 118 Referral Guidelines for imaging were published in 2000 by the European Commission (based on the RCR 1998 publication Making the Best Use of a Department of Clinical Radiology) [18]. The American College of Radiology's Appropriateness Criteria [19] and Western Australia's

Diagnostic Imaging Pathways [21] provide evidence based guidance considering global evidence. Once the criteria for referring patients to radiological examination are set, an effective way of improving good justification practices to include it as part of a programme of clinical audit [22].

Nuclear medicine refers to the practice in which unsealed radioactive substances are administered to patients for diagnosis, treatment or research. The radiation exposure comes from the radioactive substance administered to the patient. In the context of justification, the nuclear medicine specialist should select the appropriate test likely to give the expected result on the grounds of accepted current medical knowledge, taking into account the patient's dose, whether the patient is pregnant, lactating or a child, and local resources [23]. The efficacy, benefits and risks of alternative technology, for example ultrasound or magnetic resonance imaging, should be taken into account. Ideally, referring physician, should consult the nuclear medicine specialist for the appropriate procedure to be performed.

Optimization of Protection

Once examinations are justified, they must be optimized. Dealing with optimization of protection for patients is a unique challenge. In optimization of protection in medical exposures to ionizing radiation, the same person gets the benefit and suffers the risk. In addition, any individual restrictions on patient dose could be counterproductive in the context of medical goal of the procedure. The basic aim of this optimization of protection is to adjust the protection measures for a source of radiation in such a way that the net benefit is maximized [16]. Since radiological procedures are thought to carry some health risk, it is essential that x-ray imaging be performed within the framework of the established principles of radiation protection [14, 16]. Optimization of protection is one of these principles, but there are many aspects that need to be considered in its application [16, 23, 24]. As already mentioned, diagnostic and interventional radiology has been undergoing rapid changes in technology and practices that have implications on radiation safety of the patient. In many countries even if modern equipment is available, it cannot be assumed that the technology is being used appropriately and safely. Even more, in many places obsolete and older technologies and practices are still in use [6, 7, 9, 10, 25-27].

The optimization of radiological protection in medial exposure to ionizing radiation is usually applied at two levels: the design, appropriate selection, and construction of equipment and installations; and in daily practice by utilization of proper working procedures [16, 28, 29]. The optimization of radiological protection means keeping the doses "as low as reasonably achievable, economic and societal factors being taken into account", and is best described as management of the radiation dose to the patient to be commensurate with the medical purpose. Therefore, the basic aim of this optimization of protection is to adjust the protection measures for a source of radiation in such a way that the net benefit is maximized. Optimization of the protection should be both generic for the examination type and all the equipment and procedures involved. It should also be specific for the individual, and include review of whether or not it can be effectively done in a way that reduces dose for the particular patient while maintain the diagnostic goal of the examination. Both in diagnostic and interventional procedures using x-rays and in nuclear medicine imaging, acceptable image quality with the minimum patient dose should be the objective of the diagnostic process as a whole. It is important to note that the optimization of protection in medical exposures does not necessarily mean the reduction of doses to the patient. For example, diagnostic radiographic equipment often uses antiscattering grids to improve the image quality. However, the use of grid is associated with increase dose. If greed is removed, the dose will be reduced, but net benefit would be also reduced due to loss of image quality. Nevertheless, in the radiography of small children, the amount of scattered radiation is less the dose reduction can be achieved by removing the greed, without substantial loss if image quality.

Diagnostic Reference Levels

In diagnostic and interventional procedures using x-rays and diagnostic nuclear medicine, diagnostic reference levels (DRLs) are a well-established tool, used in optimization of protection and safety. Periodic assessments are to be performed of typical patient doses or, for radiopharmaceuticals, activities administered in a medical radiation facility. Doses in this context may be expressed in one of the accepted dosimetric quantities as described later in this chapter. If comparison with established diagnostic reference levels shows that the typical patient doses or activities are either unusually high or unusually low, a local review is to be initiated to ascertain whether protection and safety has been optimized and whether any corrective action is required [11, 16, 30, 31]. DRLs are not considered to be the dose limits.

In establishing DRLs it is preferable that, for common imaging procedures, typical (e.g., average or median) doses for patients are obtained from a representative sample of rooms and facilities. This sample should reflect both good and bad practice within a facility. The value of the DRL for that particular procedure is typically the rounded 75th percentile of the distribution of the room/facility typical doses [32]. In diagnostic nuclear medicine, an "optimum" value for a DRL is used: a reference level for administrations of activities of radionuclides sufficient to obtain information for standard groups of patients (adults and children), based on the experience of the professional groups (expert judgment) [32]. In establishing DRLs, it is fundamental to include only radiological procedures whose image quality is adequate for the medical purpose.

Recently, a concept of acceptable quality dose (AQD) has been proposed. This concept starts with a facility rather than national levels and thus promotes facility-based actions. It is also based on clinically acceptable image quality that is the primary goal of any imaging and covers all crucial parameters, including image quality, dose and the patient's body size [33].

DOSIMETRIC QUANTITIES USED RELATED TO DIAGNOSTIC AND INTERVENTIONAL RADIOLOGY

Diagnostic imaging generally covers a diverse range of examination types, many of which are increasing in frequency and technical complexity.

This has resulted in the development of new dosimetric measuring instruments, techniques and terminologies which affect the work both in the clinical environment and calibration facilities.

There are two fundamental reasons for measuring or estimating the patient dose in medical imaging [34]: a) to provide a means for setting and checking standards of good practice, as an aid to the optimization of the radiation protection of the patient and of image quality; and b) To estimate the absorbed dose to tissues and organs in the patient in order to assess radiation damage so that radiological techniques can be justified and cases of accidental overexposure investigated.

Dosimetric quantities used for these purposes can be divided in several categories: basic dosimetric quantities, application specific quantities and risk-related quantities. Basic dosimetric quantities are fundamental quantities based on which we define application specific quantities and risk-related quantities that are used in diagnostic radiology measurements [13, 34-39].

Basic Dosimetric Quantities

Fluence and Energy Fluence

Fluence Φ is defined as the quotient dN by da, where dN represents the number of particles incident on a sphere of cross-sectional area da:

$$\phi = \frac{dN}{da} \tag{1}$$

The unit of fluence Φ is reciprocal square meter (m^{-2}).

Energy fluence ψ is used to specify the energy carried by the photons in x-ray beam. It is the quotient of radiant energy dR incident on a sphere of cross-sectional area da:

$$\psi = \frac{dR}{da} \tag{2}$$

and is expressed in unit joule per square meter (Jm^{-2}).

The temporal change of these quantities is measured by fluence rate and energy fluence rate, as:

$$\dot{\phi} = \frac{d\phi}{dt} \tag{3}$$

$$\dot{\psi} = \frac{d\psi}{dt} \tag{4}$$

with their units being $m^{-2}s^{-1}$ and Wm^{-2}, respectively.

Fluence and energy fluence find their application in situations in which radiation interactions are not dependent of the direction of the incoming particles. In cases where differential solid angle is involved quantities such as particle radiance and energy radiance are used.

Kerma and Kerma Rate

Kerma (kinetic energy released in matter) represents the sum of the initial kinetic energies of all the charged particles liberated by uncharged particles, dE_{tr}, in a mass dm of the medium:

$$K = \frac{dE_{tr}}{dm} \tag{5}$$

Kerma unit has a special name gray (Gy), although by definition it is joules per kilogram (J/kg). For uncharged particles that have single energy we can relate kerma to the energy fluence by using mass energy transfer coefficient, μ_{tr}/ρ of the material:

$$K = \frac{\mu_{tr}}{\rho} \psi \tag{6}$$

For the application of x-rays in medical imaging we usually express kerma in air, in which case air kerma is related to the energy fluence by the mass energy coefficient for air $(\mu_{tr}/\rho)_a$:

$$K_a = \left(\frac{\mu_{tr}}{\rho}\right)_a \psi \tag{7}$$

Since x-ray spectrum used in medical imaging is polyenergetic, a mean value of $(\mu_{tr}/\rho)_a$ should be used, applying weighting factors according to the energy distribution of the energy fluence.

Kerma rate is a quotient of dK by dt, where dK is the increment of kerma in the interval of time:

$$\dot{K} = \frac{dK}{dt} \tag{8}$$

Kerma rate unit is (J/kg)/s or Gy/s.

Energy Imparted

The mean energy imparted, $\bar{\varepsilon}$ (unit: J), to the matter in a given volume equals the radiant energy, R_{in}, of all charged and uncharged ionizing particles which enter the volume minus radiant energy, R_{out}, of all those charged and uncharged particles which leave the volume, plus sum of all changes in rest energy of nuclei (not significant in diagnostic radiology):

$$\bar{\varepsilon} = R_{in} - R_{out} + \sum Q \tag{9}$$

expressed in joules (J) or also in electron-volts (eV). For the photon energies used in diagnostic radiology, ΣQ is equal to zero.

Absorbed Dose

The absorbed dose (D) is the quotient of $d\bar{\varepsilon}$ by dm, where $d\bar{\varepsilon}$ is the mean energy imparted to matter of mass dm:

$$D = \frac{d\bar{\varepsilon}}{dm} \tag{10}$$

Absorbed dose has a unit of J/kg but in practice, unit name gray (Gy) is in use.

By using mass energy coefficients, absorbed dose can be related to the air kerma. This relation is important because it allows interpreting response of air kerma detectors and dosemeters in terms of absorbed dose in material of interest (such as tissue equivalent material and water):

$$D = \frac{\left(\frac{\mu_{en}}{\rho}\right)_{mat}}{\left(\frac{\mu_{tr}}{\rho}\right)_{air}} K_{air} \tag{11}$$

In the case of charged-particle equilibrium and in the absence of bremsstrahlung losses, the kerma is numerically equal to the absorbed dose.

Application Specific Dosimetric Quantities

Set of dosimetric quantities and units used for patient dose assessment in the diagnostic and interventional radiology reflects recently achieved international harmonization in the field promoted by International Commission for International Units and Measurements (ICRU) and International Atomic Energy Agency (IAEA) [36, 27]. This concept is now based on the dosimetric quantity air kerma. A number of earlier publications have expressed measurements in terms of the absorbed dose to air. Recent publications point out the experimental difficulty in determining the absorbed dose to air, especially in the vicinity of an interface; in reality, what the dosimetry equipment registers is not the energy absorbed from the radiation by the air, but the energy transferred by the radiation to the charged particles resulting from the ionization. For these reasons, ICRU recommend the use of air kerma rather than absorbed dose to air, that applies to quantities determined in air, such as the entrance surface air kerma (rather than entrance surface air dose) and the kerma-area product (rather than dose–area product) [37].

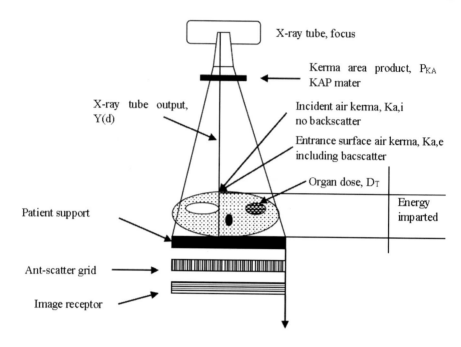

Figure 1. Measuring arrangement scheme in diagnostic radiology.

There is number of practical dosimetric quantities that are useful for measurements when it comes to the medical use of x-rays. One of the important issues is the choice of the position of the point of measurement or calculation of the quantities with the respect to the x-ray tube focal spot and the patient or phantom. Due to the fact that diverging radiation beams used in medicine are unchanging, the intensity of kerma and dose will decrease with the distance from the x-ray focal tube following the inverse-square law. Although the air kerma is constant for all points in the radiation field incident on the body, backscatter radiation from within the patient (or phantom) will have significant impact on the kerma and dose at the entrance surface (with factors up to 1.6 for general radiology). Because of this, distance from the patient of phantom must also be specified.

Since most of these quantities will be measured with instruments calibrated in terms of air kerma, it was found useful to name these quantities in terms of air kerma, unless measured or calculated inside a phantom or a patient, when absorbed dose becomes quantity of choice. Descriptive words

Radiation Exposure in Medical Imaging

are added to the names to indicate the position of the measurements and whether backscatter radiation is included or not (incident and entrance, respectively). Subscripts that define material in which the quantity is expressed is added (such as *a* for air) and the measurement condition (*i* for incident and *e* for entrance). Application specific quantities with their symbol and field of application are given in Table 1.

Table 1. Dosimetric quantities used in medical x-ray applications

Dosimetric quantity name	Symbol	Field of application
Incident air kerma	$K_{a,i}$	Radiography and fluoroscopy
Incident air kerma rate	$\dot{K}_{a,\iota}$	Fluoroscopy
Entrance surface air kerma	$K_{a,e}$	Radiography and fluoroscopy
Entrance surface air kerma rate	$\dot{K}_{a,e}$	Fluoroscopy
Air kerma-area-product	P_{KA}	Radiography and fluoroscopy
Air kerma-area-product rate	$\dot{P}_{KA}$	Radiography and fluoroscopy
Air kerma-length product	P_{KL}	Computed tomography
CT air kerma index	C_k	Computed tomography

Incident Air Kerma

The incident air kerma is the air kerma from the incident beam on central x-ray beam axis at the focal spot-to-surface distance (d_{FSD}) and it includes only primary radiation which is incident on the patient or the phantom. Unit for incident air kerma is J/kg or gray (Gy). Using inverse square law, with few corrections (due to attenuation in air, scatter in air and the structure of the x-ray source), the air kerma free-in-air at distance d can be defined as:

$$K_{a,i} = \left(\frac{d}{d_{FSD}}\right)^2 K_a(d) \tag{12}$$

The incident air kerma rate is the quotient of the increment of incident air kerma in the time interval dt and is expressed in (J/kg)/s or Gy/s:

$$\dot{K}_{a,\iota} = \frac{dK_{a,i}}{dt} \tag{13}$$

Entrance Surface Air Kerma

The entrance surface air kerma is the air kerma on the central x-ray beam axis at the point where the x-ray beam enters the patient or phantom (Figure 1). Both primary and backscattered radiation are used in measurements and calculations. Unit used for entrance surface air kerma is J/kg or its special name gray (Gy). With the use of backscatter factor B, which depends on x-ray spectrum, x-ray field size and thickness and composition of the phantom (or patient), entrance surface air kerma can be related to the incident air kerma by:

$$K_{a,e} = K_{a,i}B \tag{14}$$

The entrance surface air-kerma represents the change of entrance surface air kerma in time interval dt in units of (J/kg)/s or Gy/s:

$$\dot{K}_{a,e} = \frac{dK_{a,e}}{dt} \tag{15}$$

Air Kerma-Area Product

The air kerma-area product is the integral of the air kerma free-in-air over the area A of the x-ray beam in a plane which is perpendicular to the beam axis.

$$P_{KA} = \int_A K_a(A)dA \tag{16}$$

Unit for air kerma-area product is $J/(kg/m^{-2})$ or $Gy \cdot cm^2$. Often, symbols DAP or KAP are used for kerma-area product.

The air-kerma product has the useful property that it is almost constant over the distance from the x-ray tube focus (when interactions in air and scatter can be neglected), provided that the planes of measurement and calculation are not so close to the patient or phantom so there is no influence of backscatter radiation. The air kerma-area product rate is the quotient of dP_{KA}, which is the increment of P_{KA} in time interval dt, by dt:

$$P_{KA} = \frac{dP_{KA}}{dt} \tag{17}$$

Unit for air kerma-area product is $J \cdot m^2/(kg \cdot s)$ or $Gy \cdot cm^2/s$.

Air Kerma-Length Product

Integral of the air kerma over a line of length L parallel to the axis of the beam is denoted as air kerma-length product, in units of $(J/kg) \cdot m$ or $Gy \cdot cm$:

$$P_{KL} = \int_L K_a(L)dL \tag{18}$$

Quantities for CT Dosimetry

The CT air kerma index free-in-air (C_K) is the integral of the CT axial air kerma profile $K_a(z)$, along the axis of the rotation of the CT scanner, z, for a single rotation with a single slice divided by the nominal slice thickness T:

$$C_K = \frac{1}{T}\int_{-\infty}^{+\infty} K_a(z)dz = \frac{P_{KL}}{T} \tag{19}$$

For multi-slice scanner with N_i slices of thickness T_i, C_k becomes:

$$C_K = \frac{P_{KL}}{T_i N_i} \tag{20}$$

Similarly to CT air kerma index free-in-air, CT air kerma index in phantom, $C_{K,PMMA}$, can be defined for measurements in standard polymethylmethacrylate (PMMA) CT dosimetry phantom (which needs to be positioned centrally in the gantry of the scanner, with their longitudinal axis coinciding with the axis of the rotation or the scanner):

$$C_{K,PMMA} = \frac{1}{T}\int_{-\infty}^{+\infty} K_{a,PMMA}(z)dz = \frac{P_{KL,PMMA}}{T} \tag{21}$$

Unit for both CT air kerma free-in-air and in phantom is J/kg or gray (Gy).

The International Electrotechnical Commission and European Commissions [37, 38] recommend an integration length of 100 mm and that the kerma is expressed in air, K_a, which gives:

$$C_{K,PMMA,100} = \frac{1}{T} \int_{-50}^{+50} \frac{K_{a,PMMA}(z)dz}{T_i N_i} \tag{22}$$

In order to provide an indication of the average air kerma in phantom over single rotation, we can measure $C_{K,PMMA,100}$ in the center ($C_{K,PMMA,100,c}$) and at the periphery ($C_{K,PMMA,100,p}$) of the head and body phantoms. Denoted value is weighted CT air kerma index in phantom $C_{K,PMMA,w}$:

$$C_{K,PMMA,w} = \frac{C_{K,PMMA,100,c} + 2C_{K,PMMA,100,p}}{3} \tag{23}$$

Unit for $C_{K,PMMA,w}$ is J/kg or gray (Gy).

According to the recommendation of the European guidelines on quality criteria for CT, $C_{K,PMMA,100}$ should be determined at the center $C_{K,PMMA,100,c}$ and at 10 mm below the surface $C_{K,PMMA,100,p}$ of the standard CT dosimetry phantom and that $C_{K,PMMA,100,p}$ should be average of measurements at four locations around the periphery of the phantom.

$C_{K,PMMA,w}$ can be normalized to unit tube-current exposure-time product:

$$_nC_{K,PMMA,w} = \frac{C_{K,PMMA,w}}{P_{It}} \tag{24}$$

The CT air kema-length product in the standard CT dosimetry phantom for a complete conventional examination is defined as:

$$P_{KL,CT} = \sum_j {}_nC_{K,PMMA,w_j} T_j N_j P_{It_j} \tag{25}$$

where j represents each serial scan sequence.

In the same manner, absorbed dose is often also measured in head and body CT phantoms. In that case we define CT dose index *CTDI* from which we derive:

$$CTDI_w = \frac{CTDI_c + 2CTDI_p}{3} \tag{26}$$

$$CTDI_{vol} = \frac{CTDI_w}{p} \tag{27}$$

where p is pitch factor for helical scanning. From $CTDI_{vol}$ we can also define dose-length product as:

$$CTDI_{vol} = \frac{DLP}{L} \tag{28}$$

where L is the length of scanned region.

Risk-Related Quantities

The fundamental dosimetric quantity in radiological protection is the absorbed dose D. For low level doses, mean absorbed dose in organs or tissues in the human body can be indicator of subsequent stochastic effects while the high level doses can indicate the severity of deterministic effect [13, 14, 40]. Since measuring absorbed dose in organs and tissues of the human body is not practically possible, dosimetric quantities which can be directly measured or can be estimated from related measurements have been developed. The risk-related quantities can be obtained from practical dosimetric quantities incorporating dose-conversion coefficients.

In routine diagnostic radiology, threshold for deterministic effect should never be surpassed. There are certain procedures, like interventional radiology, in which cases of skin injure and cataract have been reported as a result of prolonged fluoroscopy time. When deterministic effect is a possibility, then absorbed dose to the organs that are heavily irradiated becomes quantity of interest ($D_{T,local}$). For the probability of stochastic effects ICRP recommends the use of average absorbed dose in a tissue or organ. ICRU [39] defines average absorbed dose as the integral of absorbed dose D_t over the mass of the tissue divided by its mass m_T:

$$D_T = \frac{1}{m_T} \int_{m_T} D_t dm_t \tag{29}$$

Table 2. Radiation weighting factor [13]

Type of radiation, R	Energy range	Weighting factor, w_R
Photons, electrons	All energies	1
Neutrons	< 10 keV	5
	10 – 100 keV	10
	100 keV – 2 MeV	20
	2 – 20 MeV	10
	> 20 Mev	5
Protons	>2 MeV	5
Alpha particles, fission fragments, heavy nuclei	-	20

Beside absorbed dose, type and energy of radiation also influences the probability of stochastic effect. It is then the ICRP recommendation that for the purpose of radiation protection, organ dose be weighted by a radiation weighting factor w_R [13]. Values of w_R have been chosen to represent the relative biological effectiveness of specific type and energy of radiation and are given in Table 2.

This weighted absorbed dose has been named equivalent dose H_T and has a special unit Sievert (1 Sv = 1 J/kg):

$$H_T = \sum_R D_{T,R} w_R \tag{30}$$

Probability of stochastic effects is also dependent on the organ or tissue which is irradiated. Because of that, ICRP has defined weighted factor for equivalent dose, based on which tissue or organ is being irradiated, named tissue weighting factor w_T. The effective dose is now defined as the sum of the weighted equivalent doses in all the tissues and organs of the body:

$$E = \sum_T H_T w_T \tag{31}$$

Radiation Exposure in Medical Imaging 21

with sum of weighting tissue factors being normalized to unity [39]:

$$\sum_T w_T = 1 \tag{32}$$

Dose Conversion Coefficients for Assessment of Organ and Tissue Doses

In diagnostic and interventional radiology, it is common practice to measure a radiation dose quantity that is then converted into organ doses and effective dose by means of conversion coefficients [34, 36]. These coefficients are defined as the ratio of the dose to a specified tissue or effective dose divided by the normalization quantity. Incident air kerma, entrance surface air kerma, air kerma-length and kerma-area product can be used as normalization quantities. Therefore, conversion coefficient, c, is used to relate a dosimetric quantity to some other quantity which can be readily measured or calculated in the clinical use:

$$c = \frac{specified\ dosimetric\ quantity}{normalization\ quantity} \tag{33}$$

When considering stochastic effects, the specific dosimetric quantity can be mean absorbed dose in organ D_T or in a specialized tissue of interest, for example glandular tissue in the breast in mammography examination. For deterministic effects, the specific quantity is absorbed dose to the heavily irradiated region such as skin or eye lens in interventional radiology. If specified quantity is the organ dose, D_T, and the normalization quantity incident air kerma, entrance surface air kerma or kerma-area product, the conversion coefficient is:

$$C_{T,K_{a,i}} = \frac{D_T}{K_{a,i}} \tag{34}$$

$$C_{T,K_{a,e}} = \frac{D_T}{K_{a,e}} \tag{35}$$

$$C_{T,P_{KA}} = \frac{D_T}{P_{KA}} \tag{36}$$

with units of (J/kg)/(J/kg) or Gy/Gy, (J/kg)/(J/kg) or Gy/Gy and (J/kg)/(Jcm^2/kg) or Gy/(Gycm^2), respectively. In mammography, we use mean glandular dose, D_G, as specified quantity and air kerma as normalization quantity in terms of (J/kg)/(J/kg) or Gy/Gy:

$$C_{G,K_{a,i}} = \frac{D_G}{K_{a,i}}$$

(37)

For CT examinations, organ dose D_T can be used as specified quantity which can be related to CT air kerma index in units (J/kg)/(J/kg) or Gy/Gy:

$$C_{T,C_K} = \frac{D_T}{C_K}$$

(38)

Although effective dose is used as a principal protection quantity, it should not be used to assess risks of stochastic effects in retrospective situations for exposures in identified individuals, nor should it be used in epidemiological evaluations of human exposure. Such risks for stochastic effects are dependent on age and sex. The age and sex distributions of workers and the general population can be different from the overall age and sex distribution for the population undergoing medical procedures using ionizing radiation, and will also differ from one type of medical procedure to another. For these reasons, risk assessment for medical uses of ionizing radiation is best evaluated using appropriate risk values for the individual tissues at risk, and for the age and sex distribution of the individuals undergoing the medical procedures. Effective dose can be of practical value for comparing the relative doses related to stochastic effects from different diagnostic examinations and interventional procedures, from the use of similar technologies and procedures in different hospitals and countries and from the use of different technologies for the same medical examination [16].

DOSIMETERS USED IN DIAGNOSTIC RADIOLOGY

Ionization chambers are the main devices used for dosimetry in diagnostic radiology [34]. In the energy region considered, the free air chamber is the primary standard for realizing the unit of air kerma [34, 36]. The advantage of ionizing chamber as a dosimetric device is precision, easy use and few other complicating factors [36, 37]. Parallel-plate ionization chambers are mainly used, but cylindrical chambers are used, as well. Air kerma area-product meters are special types of parallel plate ionization chambers used to measure the integral of air kerma over the beam area. Parallel plate ionizing chambers are calibrated with plates perpendicularly oriented to the beam axis. They are also used in the same orientation for patient dose measurement. The response of the cylindrical chamber is symmetrical with respect to the chamber axis. A version of cylindrical ionizing chamber, designed for non-uniform response, is used for computed tomography dosimetry.

Other devices with special properties like thermoluminescent or semiconductor detectors are also used in diagnostic radiology. Real-time measurements can be accomplished with semiconductor dosimeters, while small size thermoluminescent dosimeters (TLDs) are used for measurements on patients. The main disadvantage of these dosimeters is their energy dependence of response that differs from ionizing chambers. In addition, those dosimeters must always be calibrated against ionizing chamber. The required accuracy and precision of a given measurement will depend on the purpose for the measurement and the type of equipment being monitored. It is recommended that the combined uncertainties of in-beam dose measurements not exceed ±10% [36].

CALIBRATION PROCEDURES FOR DIAGNOSTIC AND INTERVENTIONAL RADIOLOGY

The International Measurements System (IMS) for radiation metrology provides the framework for dosimetry in diagnostic radiology. It ensures consistency in radiation dosimetry by disseminating to users calibrated radiation instruments, which are traceable to primary standards. The IMS consists of *Bureau International des Poids et Mesures* (BIPM), national Primary Standard Dosimetry Laboratories (PSDL), Secondary Standards Dosimetry Laboratories (SSDL) and various users performing measurements [36]. A PSDL is a national laboratory designated by the government for the purpose of developing, maintaining and improving primary standards in radiation dosimetry. A PSDL participates in the international measurement system by making comparisons through the medium of the BIPM and provides calibration services for secondary standard instruments. An SSDL may be either national or regional. A national SSDL is a laboratory which has been designated by the competent national authorities to undertake the duties of a calibrating laboratory within that country. An SSDL is equipped with secondary standards which are calibrated against the primary standards of laboratories participating in the IMS [41-43].

A decade ago, the SSDL were focused only on the calibrations in the field of radiotherapy and radiation protection [44, 45], while diagnostic radiology calibrations have drawn attention in the last decade due to increased demands for establishment of quality assurance programme in diagnostic radiology [41, 42].

Calibration is an essential part of any dose measurement, in particular if these activities are related to human health. Thus, all instruments used in conventional diagnostic radiology, interventional radiology, mammography and CT must be calibrated, having a valid calibration certificate from an accredited calibration laboratory, typically SSDL. Requirements for calibration facilities, in particular for the SSDL are given in terms of necessary equipment for generation of beam qualities, dosimetry and auxiliary equipment necessary for operation of SSDL. All equipment used for calibration at an SSDL shall be of a reference class and be available in duplicate at the SSDL. This includes: ionization chambers, electrometers, thermometers, barometers and a device to measure the relative humidity of air [46]. As already mentioned, reference standard for diagnostic radiology calibrations is an ionization chamber, and it should comply with International Electrotechnical Commission (IEC) 61674 standard in order to perform measurements with sufficient accuracy and reliability [47]. Radiation beam quality is the indication of photon fluence spectrum. In practice, it is determined by the tube voltage, first and second half-vale layer (HVL) and total filtration [36]. Required radiation qualities shall be established in accordance with recommendations given in the standards of IEC [48]. The qualities used for the calibration of dosimeters for different applications are shown in the Table 3.

Table 3. Radiation qualities used for calibrations in diagnostic radiology. Adopted from [36]

Radiation quality	Radiation origin	Phantom material	Application
RQR	Unfiltered beam emerging from x-ray assembly	No phantom	General radiography, fluoroscopy, dental radiology
RQA	Radiation beam from an added filter	Aluminum	Measurements behind the patient (on the image intensifier)
RQT	Radiation beam from an added filter	Copper	CT applications (free in air)
RQR-M	Unfiltered beam emerging from x-ray assembly	No phantom	Mammography (free in air)
RQA-M	Radiation beam from an added filter	Aluminum	Measurements behind the patient (on the image intensifier)

For conventional radiography, fluoroscopy, CT and dental applications a tungsten anode tube and x-ray unit operating at the x-ray tube voltage ranging from 50 kV to 150 kV are used. For the calibration of mammography dosimeters, a molybdenum anode tube with molybdenum filtration is recommended.

The detail description of methods for establishment of beam qualities used for calibrations in diagnostic radiology are given in the International Code of Practice [36].

The general principles for the calibration of dosimeters used in diagnostic and interventional radiology are similar to principles valid for instruments used in radiotherapy and radiation protection [44, 45]. The SSDL shall provide a calibration coefficient in terms of air kerma or air kerma-length product, where appropriate. Air kerma-area product meters require great care in their calibration, as their performance depends on the actual set-up in the hospital. They may be calibrated *in situ*.

The calibration of dosimeters is usually performed using substitution method when quantity under question is measured in the same reference point by reference standard and instrument to be calibrated (user's instrument) [36, 44, 45]. The reading of the reference dosimeter is converted to air kerma or air kerma rate by means of the calibration coefficient that should be supplied by PSDL. To ensure the accuracy of measurements, the calibration and the energy dependence of response must be known to the SSDL. The calibration coefficient should have a form appropriate to the nature of the reference dosimeter.

As the output of the x-ray unit may vary with time it is recommended to use monitor chamber to check stability of the output during calibration. The indications of reference instrument and user's instrument M_{ref} and M_{user} shall be related to monitor chamber readings. For the reference radiation qualities, the calibration coefficient is given by equation:

$$N_{K,Qo}^{user} = N_{K,Qo}^{ref} \cdot \frac{M^{ref} k_{TP}^{ref}}{M^{user} k_{TP}^{user}} \cdot \frac{(mk_{TP})^{user}}{(mk_{TP})^{ref}} \tag{39}$$

where m is reading from the monitor chamber and k_{TP} correction for temperature and pressure. For other beam qualities further correction is applied [36]. The general procedure described above applies to calibrations in terms or air kerma, however, whereas calibrations in terms of kerma-area product and kerma-length product are considered as special cases. Computed tomography dosimeters are designed for non-uniform exposure from a single scan. Primary beam is not

Radiation Exposure in Medical Imaging 27

more than 10% of the full length of the chamber [49]. The calibration of this type of dosimeter is performed in air in a uniform x-ray field with known air kerma rate by irradiation of a well-defined fraction of the useful volume of the chamber. The calibrated quantity is the air kerma-length.

The actual measurement using KAP meter is the integral of the exposure over the area of the x-ray beam. The calibration of such a device needs to include the ionization response and the correct beam area. The reading is the product of air kerma and the area of the x-ray field. Various methods for KAP meter calibration are described in the literature, performed in situ and at a SSDL [36]. Larsson et al. suggested laboratory calibration by mapping X-ray field to account for field heterogeneity [50]. Such calibration may be time consuming and inappropriate for field application. Field calibration is performed in the geometry and beam quality used clinically. The air kerma-product, P_{KA}, should be determined for the x-ray beam transmitted through the chamber and incident on the patient.

The uncertainty of calibration procedure is related to the complexity and quality of the established reference radiation for a particular calibration. The contributing factors to the uncertainty budget include properties of high-quality specialized equipment, physical and radiological environment, and presence of skilled and experienced personnel. The level of uncertainty should be appropriate to the use of the measurement, and it should be derived by the each calibration laboratory. Typically, uncertainty of calibration in the field of diagnostic radiology are of order of magnitude of few percent, expressed as relative, combined and expanded uncertainty.

CLINICAL DOSIMETRY AND TYPICAL DOSE LEVELS

As already described, medical exposures include the exposure of patients as part of their medical diagnosis or treatment, the exposure of individuals as part of health screening programmes, the exposure of healthy individuals or patients voluntarily participating in medical, biomedical, diagnostic or therapeutic research programmes [13]. There are three general categories of medical practice involving exposure to ionizing radiation:

diagnostic radiology (including interventional procedures), nuclear medicine and radiation therapy.

Due to increasing importance of radiation burden for medical x-ray examination, clinical dosimetry is becoming an active research and practical area. Objectives of clinical dose measurements in diagnostic and interventional radiology are multiple, as assessment of equipment performance, optimization of practice trough establishment of diagnostic reference levels (DRL) or assessment of risk emerging from use of ionizing radiation [16, 32]. Various dosimetric quantities are needed to assess radiation exposures to humans in a quantitative way, in order to assess dose–response relationships for health effects of ionizing radiation which provide the basis for setting protection standards as well as for quantification of exposure levels. However, risk-related dosimetric quantities as absorbed dose or equivalent dose to the tissue or organ have been established. Either for measurements on patients or on phantoms, different application-specific dosimetric quantities were established. Therefore, from the clinical point of view, procedures to assess dose to standard dosimetry phantoms and patients in clinical diverse modalities, as radiography, fluoroscopy, mammography and computed tomography include measurements of: a) incident air kerma, entrance surface air kerma and kerma-area product (radiography); b) kerma-area product and entrance surface air kerma rate (fluoroscopy); c incident and entrance surface air kerma (mammography); and d) kerma-length product (computed tomography). Measurements on phantoms cannot provide an estimate of the average dose for patient population. Therefore, these must be supplemented by measurements on patients.

Diagnostic radiology generally refers to the analysis of images obtained using x-rays. These include plain radiographs (e.g., chest x-rays), images of the breast (i.e., mammography), images obtained using fluoroscopy (e.g., with a barium meal or barium enema) and images obtained by devices using computerized reconstruction techniques such as computed tomography (CT). In addition to their use for diagnosis, interventional or invasive procedures are also performed in hospitals (e.g., placing a catheter in a blood vessel to obtain images). Dental radiology has been included in the analysis conducted of diagnostic radiology practice.

Nuclear medicine refers to the introduction of unsealed radioactive materials into the body, most commonly to obtain images that provide information on either structure or organ function. The radioactive material is usually given intravenously, orally or by inhalation. A radionuclide is usually modified to form a radiopharmaceutical that will be distributed in the body according to physical or chemical characteristics (for example, a radionuclide modified as a phosphate will localize in the bone, making a bone scan possible). Radiation emitted from the body is analyzed to produce diagnostic images. Less commonly, unsealed radionuclides are administered to treat certain diseases (most frequently hyperthyroidism and thyroid cancer). There is a clear trend towards increased therapeutic applications in modern nuclear medicine [1, 51].

Clinical Dosimetry in Diagnostic Radiology

Exposures resulting from radiological procedures constitute the largest part of the population exposure from artificial radiation. There is a need to control these doses and therefore to optimize the design and use of x-ray imaging systems. It is generally recognized that even a 10% reduction in patient dose is a worthwhile objective for optimization. The main aim of patient dosimetry with respect to x-rays used in medical imaging is to determine dosimetric quantities for the establishment and use of guidance levels (diagnostic reference levels) and for comparative risk assessment. In the latter case, the average dose to the organs and tissues at risk should be assessed. Only a limited number of measurements serve for potential risk assessment of the examination or intervention. An additional objective of dosimetry in diagnostic and interventional radiology is the assessment of equipment performance as a part of the quality assurance process.

In many situations, it is of interest to make measurements directly on the patient. However, it is preferable to make measurements using a standard phantom to simulate the patient for the control of technical parameters, for the comparison of different systems and for optimization. In some cases, specialized dosimeters are required, the design and performance of which must be matched to the needs of the clinical measurement. The use of such dosimeters and/or the interpretation of the results obtained may require specialized techniques and knowledge. Various examination techniques are used in x-ray imaging and, therefore, it is essential to standardize the procedures for the dose measurement in the clinical environment. IAEA TRS 457 Code of practice contains detailed description of clinical dosimetry methods for different imaging modalities [36]. These methods are presented in Table 4.

Radiography

The principal quantities for patient dosimetry in general radiography are the incident air kerma, the entrance surface air kerma and the air kerma–area product. One or more of these quantities may be determined depending upon user requirements. For each patient, the incident air kerma can be determined by calculation from recorded exposure parameters and the measured tube output. The entrance surface air kerma provides direct assessment of patient dose and can be measured by TLDs. If the x-ray machine is equipped with a KAP meter, the value of the air kerma–area product can be recorded. When estimating the patient dose in general radiography it is useful to have a detailed information for patient examination. It provides comprehensive information about the x-ray equipment, the examination procedure and the patient, and thus helps the interpretation of the assessed doses [15, 26, 36].

The assessment of incident air kerma and the entrance surface air kerma may be achieved by measurement of tube radiation output in units mGy/mAs at a given point (without a patient) using an ionization chamber, followed by calculation of the entrance surface air kerma from recorded exposure and geometric data, as well as the use of an appropriate backscatter factor. Entrance surface air kerma may also be measured using TLDs. The dosimeters are packaged in plastic sleeves that are sterilizable, and are attached to the patient's skin using surgical tape. Correction factors for the energy dependence of the dosimeters and their sensitivity are applied to the raw TLD data. A background correction is also applied.

Radiation Exposure in Medical Imaging

Table 4. Clinical dosimetry methods. Adopted from [36]

Modality	Measurement subject	Measured quantity	Remark
General radiography	Phantom	Incident air kerma	Methodology for using chest and abdomen/lumbar spine phantoms is described
	Patient	Incident air kerma	Measurements on patient's skin
		Air kerma–area product	Calculated from exposure parameters and measured tube output
		Entrance surface air kerma	Methodology same as for fluoroscopy
Fluoroscopy	Phantom	Entrance surface air kerma rate	Measured directly on a phantom or calculated from the incident air kerma rate using backscatter factors
	Patient	Air kerma–area product	-
Mammography	Phantom	Incident air kerma	Mean glandular dose is the primary quantity of interest. It is calculated from measured incident air kerma
		Entrance surface air kerma	When this is measured (using TLDs) the backscatter factors are used to calculate the incident air kerma
	Patient	Incident air kerma	Mean glandular dose is the primary quantity of interest. It is calculated from the incident air kerma estimated from measurements of tube output by using the exposure parameters for the examination
Modality	Measurement subject	Measured quantity	Remark
CT	Phantom	CT air kerma indices	Measurements in air or in PMMA head and body phantoms
	Patient	CT air kerma–length product	Direct measurements on patients are not described in this Code of Practice. Instead, a CT air kerma–length product is calculated from patient exposure parameters and results of phantom measurements
Dental radiography	Patient	Incident air kerma	Calculated from exposure parameters and measured tube output for bitewing projection
		Air kerma–length product	Used for calculation of the air kerma–area product for a panoramic projection

Physical phantoms that simulate patient anatomy can be used for patients dosimetry. The ICRU has described the requirements for physical dosimetry phantoms [52]. Some phantoms have a fair degree of anatomical accuracy and are a reasonably accurate representation of human anatomy, both in terms of the size and position of the organs and with respect to the attenuation properties. A problem with some anthropomorphic phantoms is that they are not tissue equivalent, which leads to inaccurate dosimetry for diagnostic radiology [53]. There are limitations regarding measurements in a physical dosimetry phantom. These relate to the need to use a large number of dosimeters to estimate the dose to physically large organs, the non-uniform distribution of radiation within the phantom and the effect of small uncertainties in the position of the radiation field. As a consequence, this method of patient dosimetry as well as the other methods (measuring entrance surface air kerma using TLDs) are not suitable for routine patient dose assessments [1].

Monte Carlo computational techniques are also used to estimate organ or tissue doses. These are computer-based methods that employ computational models to simulate the physical processes associated with the interaction of an x-ray beam with the human body. There are two types of computational model: mathematical and voxel phantoms [54, 55]. Mathematical phantoms are a three-dimensional representation of a patient. The organs and the whole body are defined as geometric bodies (such as cylinders and ellipsoids). Voxel phantoms are based on either CT or MRI images of actual patients. Organ sizes and positions are deduced from the volume elements determined from the imaging data.

As a consequence these phantoms are physically more accurate, the only limitation being the size of the voxels used. Monte Carlo calculations are used to deduce energy deposition of x-ray photons in computational models of human anatomy [34]. Normally, patient dose is assessed by applying suitable Monte Carlo calculated conversion coefficients to a routinely measured quantity such as kerma area product or entrance surface air kerma. As already described, if the air kerma at a specified point is known, it is possible to use normalized organ dose data to deduce organ doses for a typical patient, effective dose being calculated from the organ doses.

Normalized organ dose data are available for many examination types generally based upon Monte Carlo simulations of examinations [56].

Table 5. Radiation dose levels in planar radiography [1, 57-60]

Procedure	Projection	$K_{a,e}$ [mGy]	E [mSv]
Chest	PA	0.02 – 5.06	0.02 – 0.14
	LAT	0.20 – 13.74	0.02 – 0.37
Head	-	0.07 – 25.80	0.01 – 0.40
Abdomen	-	0.70 – 16.70	0.25 – 3.62
Pelvis	-	0.70 – 15.60	0.20 – 2.68
Spine	Lumbar AP/PA	1.30 – 17.40	0.27 – 2.20
	Lumbar LAT	2.13 – 37.40	0.22 – 3.30
	Thoracic AP	0.70 – 15.50	0.18 – 1.43
	Thoracic LAT	1.64 – 26.90	0.16 – 2.90
	Cervical AP	0.25 – 6.90	0.02 – 0.40
	Cervical LAT	0.20 – 7.20	<0.01 – 0.31

Numerous investigations were performed in order to estimate the radiation dose level in planar radiography [1, 57-60]. The dose level for most frequent examination types (head, chest, abdomen, pelvis and spine) are presented in Table 5.

Fluoroscopy

Approaches to patient dosimetry are different for procedures that involve the use of fluoroscopy equipment [61]. During such examinations an automatic brightens control is used to adjust the generator settings to compensate for changes in attenuation in the x-ray beam and to keep image quality constant. Consequently, the tube potential and tube current change continuously as the projection direction changes because of changes in attenuation through the patient. Furthermore, the anatomical area of the patient irradiated by the primary beam varies, and different tissues have different attenuation coefficients. This means that it is difficult to monitor maximum entrance surface air kerma directly, as the anatomical position where this occurs may not be known in advance [62]. In addition, dosimeters placed on the patient's skin may not be in the primary beam for all projection directions used in some procedures (e.g., interventional cardiology). In these circumstances, kerma–area product (KAP or $P_{K,A}$) may be assessed, using a transmission ionization chamber. It must be calibrated in situ, because for geometry involving an undercouch x-ray tube and overcouch detector the attenuation of the patient couch must be taken into account [63]. The uncertainty of KAP readings is approximately 6% for an overcouch X-ray tube geometry [50] and up to 20% for an undercouch x-ray tube geometry, depending on how well the dosemeter has been calibrated [63].

Organ doses resulting from fluoroscopy procedures may also be assessed using TLDs loaded into a physical phantom. Dosimeters may be placed in the phantom at positions corresponding to the organs of interest, and a typical fluoroscopy procedure is simulated on the phantom using the appropriate x-ray equipment [64].

There is increasing concern about skin dose levels in cardiology and interventional radiology because of the discovery of deterministic injuries in patients who have undergone long procedures using suboptimal equipment and performed by individuals inadequately trained in radiation protection [1, 27]. Thus in order to avoid serious injuries the practical actions to control dose need to be implemented in clinical practice. The ICRP in Publication 85 recommends the evaluation of doses absorbed by patient in the area that receives maximum dose [65].

A few methods have been proposed to measure local maximum skin dose (MSD) from dose distribution. The most frequently discussed in scientific literature are large, TL dosemeters arrays, large-area slow films or self-developing Gafchromic films [66, 67]. Although, the accuracy of large film methods in determining local maximum skin dose might be good, at least theoretically, one has to take into account not only their suitability but also costs (which are rather considerable) and practicability (they are laborious) when planes to implement them in routine dosimetry. Due to the latter the measurements with films or TLD arrays are rather not likely to become preferable method to monitor the local maximum dose in hospitals [67]. The availability of direct estimation/evaluation of local maximum dose from beam monitoring quantities is desirable for prevention of deterministic injuries. Depending on the x-ray unit there are various beam monitoring quantities displayed by the system like kerma-area product or/and cumulative dose. Sometimes, additional ionizing chamber is installed on x-ray tube for measurements of beam air kerma at the reference point distance, specified by operator. The correlation between the beam monitoring quantities and local MSD will depend on the concentration of the dose which in turn depends on projection and collimator settings used by the operators as well as on the distance at which the beam monitoring quantity of interest is specified. Thus the relations between them defined in one hospital usually are not applicable in another one [68].

Table 6. Radiation dose level in fluoroscopy and fluoroscopically guided (interventional) procedures [1, 30, 69-73]

Procedure		$P_{K,A}$ [Gy·cm^2]	E [mSv]
Interventional	PTCA	2.4 – 818	5.7 – 22
	Cerebral	50 – 77	3.3 – 6
	Vascular	6 – 170	7 – 16
	Others	0.5 – 99.3	9 – 26
Angiography	Non-cardiac	16 – 88	0.32 – 15
	Cardiac	1 – 639	5 – 17
Diagnostic	Upper GI	9 – 32	0.31 – 13
	Lower GI	19 – 89	0.40 – 14

Measurement of kerma-area product is likely a method of choice for assessing the doses and effective dose, and hence the potential risks, resulting from interventional procedures. Kerma-area product correlates reasonably well with radiation risk by means of conversion factors [56]. These conversion factors are examination-specific and may be deduced from Monte Carlo organ dose calculations made for simulated interventional procedures.

Data on radiation doses in fluoroscopy and fluoroscopically (interventional) procedures based on current literature are given in Table 6.

Computed Tomography

The dosimetry IN CT is based upon the use of three dedicated quantities: weighted CT dose index ($CTDI_w$), volume-weighted CT dose index ($CTDI_{vol}$) and dose–length product (DLP). Dedicated CT dosimetry phantoms are also recommended by the ICRU [52]. The phantom is placed on the CT scanner couch so that the scanner's axis of rotation coincides with the longitudinal axis of the phantom. The center of the CT scanner slice or multiple slices is aligned to the center of the phantom. Measurements are made at the center and periphery of the CT dosimetry phantom made of PMMA. CT dosimetry is based upon the use of PMMA phantoms with diameters of 16 cm and 32 cm to represent an adult head and body, respectively. Measurements are made, usually with a pencil ionization chamber of 100 mm length, at the center of the phantom and 1 cm below the surface at four equally spaced locations.

The CT dosimetry formalism based on the CT dose index is the most commonly used today. However, the International Atomic Energy Agency (IAEA) have recently adopted the terminology detailed by the International Commission on Radiation Units and Measurement (ICRU) in Report 74 (ICRU 2005), which is already described in this chapter. This approach uses more precise definitions based on air kerma, rather than on absorbed dose [34, 36]. Therefore, the basic quantities recommended to be used for dosimetry in CT are the CT air kerma indices, $C_{a,100}$ and C_W. A further CT air kerma index, C_{VOL}, is derived from C_W for particular patient scan parameters. Patient doses for a complete examination are described in terms of the CT air kerma–length product, $P_{KL,CT}$.

One of the problems associated with performing patient dosimetry measurements using CTDI on CT scanners with a large number of rows of detectors is the required integration length. For a nominal beam width of 128 mm, an integration length of 300 mm is required if scattered radiation is to be appropriately assessed. Conversion factors have been developed to allow a standard CTDI phantom and a 100-mm-long ionization chamber to assess CTDI on multislice CT scanners [74].

Effective dose E may be inferred from the DLP using appropriate conversion coefficients (E_{DLP}). Conversion coefficients have been calculated

Radiation Exposure in Medical Imaging

for different regions of the body at a range of standard ages and paediatric patients [75]. These conversion coefficients are derived from mathematical phantoms using Monte Carlo modelling and from a series of measurements made using anthropomorphic phantoms that simulate a range of ages from 0 to 15 years, into which TLD_s had been placed.

The IEC has recognized the necessity to display the dosimetric information on CT scanners. Therefore, $CTDI_{vol}$ and DLP are commonly available from the CT units [37]. The IEC has also considered developing a standard for the recording of dosimetry data in the DICOM header.

Radiation dose level in CT is usually expressed in terms of $CTDI_{vol}$ and DLP values. However, in order to enable better comparison between different scanner models and type of CT examinations radiation dose can be presented also in terms of effective dose. The typical dose levels for different CT examinations are presented in Table 7.

**Table 7. Radiation dose level in Computed Tomography (CT)
[1, 75-79]**

Procedure	$CTDI_{vol}$ [mGy]	DLP [mGy·cm]	E [mSv]
Head	34 – 1200	183-2173	0.81 – 7.8
Thorax	0.1 – 400	50-2157	0.1 – 11
Abdomen	3.5 – 800	58-2537	3.7 – 21
Spine	8.7 – 372	47-800	1.1 – 19
Pelvis	10 – 451	67-1984	6.9 – 11

Mammography

Mammography imaging is highly specific as breast tissue has a relatively high sensitivity to some adverse effects of radiation, whereas mammography requires a higher exposure to produce the required image quality. The higher exposure, compared to other radiographic procedures, is because the breast is a soft tissue organ and has very low contrast. In mammography, the objective is to produce images that provide maximum visualization of breast anatomy and the signs of disease without subjecting the patient to unnecessary radiation. As low-energy x-rays are used, dosimetry in mammography is particularly difficult. This places particular demands on the instruments used to measure breast dose, as they need to be energy independent down to 15 keV and an appropriate calibration factor should be applied. Mean glandular dose can be assessed based on measurement on phantoms or measurement on patients [34, 36]. In general, the dose to the breast is determined by a combination of three factors related to the equipment being used, to the technique factors selected for the examination, and to the size and density of the patient's breasts. The major factors determining this dose are the sensitivity of the receptor (film- screen combination, characteristics of digital receptors), and the setting of the automatic exposure control (AEC) level to produce a specific film density or other image quality factor. The dose generally increases with increased breast size and density, for a given image quality index.

It is widely acknowledged that within the breast it is the glandular tissue that is most radiosensitive, rather than fat or connective tissue. Mean glandular dose (MGD)) has been recommended by the ICRP as the relevant dosimetric quantity for mammography [34]. While the MGD correlates reasonably well with the associated radiation risk, it cannot be measured directly and therefore has to be inferred from other measurements. Mean glandular dose, MGD, can also be derived from incident air kerma, $K_{a,i}$, to a standard breast phantom. A conversion coefficient is used to deduce MGD (ICRU 2005):

$$D_G = c_G\, K_{a,i} \;(\text{Gy}) \tag{40}$$

Radiation dose levels for clinical and screening mammography are presented in Table 8.

Table 8. Radiation dose level in Mammography [1, 80-82]

Mammography	MGD [mGy]	E [mSv]
Screening	0.47 – 6.07	0.10 – 1
Clinical	1.1 – 44.80	0.13 – 1.20

Dental Radiography

In dental radiography, exposure settings are normally fixed and do not vary from patient to patient although different protocols may be used for the different types of teeth of adults and pediatric patients. Patient exposures are therefore surveyed using measurements free in air. In intraoral radiography, incident air kerma, $K_{a,i}$, is the most common used quantity for clinical dosimetry. This can be converted to the entrance surface air kerma by multiplication with a suitable backscatter factor. The air kerma–area product, P_{KA}, can be obtained by multiplying $K_{a,i}$ by the beam area A, which is well defined by the spacer/director cone. For panoramic examinations, the air kerma–area product, P_{KA}, has been adopted as the quantity for patient dose measurements. It is obtained from the measured air kerma–length product, P_{KL}, multiplied by the height of the x-ray beam, H. The P_{KL} is the integral of the free in air profile of the air kerma across the front side of the slit of the secondary collimator [36].

Phantom measurements in dental radiography are usually performed on anthropomorphic phantoms in order to derive organ doses and thereby determine the effective dose and/or the energy imparted to the patient [34, 36]. Measurements in anthropomorphic phantoms are performed using TLDs positioned in drilled holes in the phantom. Radiation dose level in dental radiography, excluding the cone beam CT examinations, is presented in Table 9.

Table 9. Radiation dose level from Dental Imaging [1, 80, 83, 84]

Procedure	ESAK [mGy]	E_{ff} [mSv]
Intraoral	0.30 – 16.09	<0.01 – 0.10
Panoramic	1.5 – 3.9	0.01 – 0.15
	CTDI [mGy]	
CT	12	0.01 - 4

Cone Beam CT

Cone Beam Computed Tomography (CBCT) represents an emerging technology that enables high-resolution volumetric scanning of the anatomy under investigation. Just as in MDCT, use of CBCT is steadily increasing in clinical practice. Even though it is a relatively new modality, CBCT is already being used for a variety of clinical applications, such vascular and non-vascular interventional procedures, dental examinations, orthopedic surgery and intraoperative procedures in general, neuroradiology, and urology and needle interventions [85-93].

This modality can be defined as hybrid system of conventional CT and angiographic unit and therefore, one can be confused when using such a system in terms of patient doses and dosimetry in general. Many investigations shown the advance of CBCT used instead of conventional CT. CBCT systems usually have superior spatial resolution for high-contrast objects but inferior contrast resolution for low-contrast objects. Because of the cone-beam nature of the irradiated field and the associated non uniformities in the primary and scatter radiation imparted to the scan volume, the standard dose metrics used in MDCT cannot be applied to CBCT [85]. Furthermore, CBCT technology is relatively new imaging modality and unique clinical dosimetry methods are not established yet. Data on radiation dose levels for this modality are presented in Table 10.

Table 10. Patient doses in cone beam computed tomography (CBCT) [85-93]

Procedure	Indication	Quantity	Value
Non-vascular interventional procedures	Liver intervention, abscess drainage, skeletal interventions	CTDI E	(1.9-23) mGy (4.43 - 18.79) mSv
Vascular head/body interventions	Tumor embolization, bleeding, revascularization in peripheral occlusive disease	Head: CTDI E	(9-75) mGy (2.1-11.5) mSv
Vascular cardiac interventions	Electrophysiological catheter ablation	E	7.9 mSv
Orthopedic interventions	Osteosynthesis	$CTDI_w$	(1.2 – 16.6) mGy
ENT and head	Paranasal sinus, temporal bone	CTDI E	(0.9 –11) mGy (0.09 – 0.9) mSv
Dental and maxillofacial	Dental workup	E	(0.1-916) µSv

Clinical Dosimetry in Nuclear Medicine

In nuclear medicine for patient dose estimation the internal radiation dose has to be estimated. The MIRD (medical internal radiation dose) system was developed primarily for use in estimating radiation doses received by patients from administered radiopharmaceuticals [1, 94]. It is based on the following equation:

$$D = N \cdot DF \tag{41}$$

where N is number of disintegrations that occur in the source organ and DF is given by the following equation:

$$DF = \frac{k \sum_i n_i E_i \phi_i}{m} \tag{42}$$

where n_i is number of particles with energy E_i emitted per disintegration, E_i energy of emitted particle, Φ_i fraction of energy emitted that is absorbed in the target, m is mass of the target region and k is constant that is used to resolve the units $(\text{Gy} \cdot \text{kg} \cdot (\text{MBq} \cdot \text{s} \cdot \text{MeV})^{-1})$.

The equation for absorbed dose in the MIRD system is:

$$D_{r_k} = \sum_h \widetilde{A_h} \, S(r_k \leftarrow r_h) \tag{43}$$

where r_k represents the region of target and r_h represents the region of source. The term $\tilde{A}_h$ represents the number of disintegrations in a source region h. All other terms are contained in the factor S:

$$S(r_k \leftarrow r_h) = \frac{k \sum_i n_i E_i \phi_i (r_k \leftarrow r_h)}{m_{r_k}} \tag{44}$$

The ICRP has developed a system for calculating internal doses to radiation workers who inhale or ingest radionuclides. However, the ICRP has also published extensive compendia of dose estimates for radiopharmaceuticals supporting the design of a kinetic model for each of the (over 100) radiopharmaceuticals, as well as dose estimates for adult and 15-, 10-, 5- and 1-year-old subjects [95, 96].

The effective dose is the quantity used for the purpose of gauging stochastic risks from radiation exposure [14, 16]. The discussion above concerning the limitations of the use of effective dose for assessing the exposures due to medical radiology also apply to its use for assessing exposures due to nuclear medicine. Thus, although the quantity has limitations, it is used here as a surrogate to assess patient exposures because of its convenience. In Table 11 the activity per procedure in MBq, coefficients mSv/MBq and effective dose (E) are given for different procedures and radiopharmaceuticals used in nuclear medicine.

Table 11. Effective doses from typical nuclear medicine procedures (Adults) [1, 94-102]

Procedure	Effective dose per administrated activity [mSv/MBq]	Administrated activity [MBq]	E [mSv]
^{11}C raclopride	5.00×10^{-3}	369 - 429	1.845 – 2.145
^{14}C urea (normal)	3.10×10^{-2}	0.037	1.15×10^{-3}
^{14}C urea (Heliobacter positive)	8.10×10^{-2}	0.037	3.00×10^{-3}
^{57}Co cyanocobalamin (IV, no carrier)	4.40×10^{0}	0.037	1.63×10^{-1}
^{57}Co cyanocobalamin (IV, with carrier)	4.60×10^{-1}	0.037	1.7×10^{-2}
^{57}Co cyanocobalamin (oral, no flushing)	3.10×10^{0}	0.037	1.15×10^{-1}
^{57}Co-7 cyanocobalamin (oral, with flushing)	2.10×10^{0}	0.037	7.77×10^{-2}
^{51}Cr sodium chromate RBCs	1.70×10^{-1}	5.6	9.5×10^{-1}
^{18}F FDG	1.90×10^{-2}	370	7.0×10^{0}
^{18}F FET	1.60×10^{-2}	180	2.88×10^{0}
^{18}F FLT	1.50×10^{-2}	3.7 x 1/kg	-

Radiation Exposure in Medical Imaging

Procedure	Effective dose per administered activity [mSv/MBq]	Administrated activity [MBq]	E [mSv]
^{18}F choline	2.00×10^{-2}	240 - 340	$(4.8 - 6.8) \times 10^{0}$
^{18}F fluoride	1.70×10^{-2}	2 x 1/kg	-
^{67}Ga citrate	1.00×10^{-1}	185	1.85×10^{1}
^{123}I hippuran	1.20×10^{-2}	14.8	1.78×10^{-1}
^{123}I MIBG	1.30×10^{-2}	14.8	1.92×10^{-1}
^{123}I sodium iodide (0% uptake)	1.10×10^{-2}	14.8	1.63×10^{-1}
^{123}I sodium iodide (35% uptake)	2.20×10^{-1}	14.8	3.26×10^{0}
^{125}I albumin	2.20×10^{-1}	0.74	1.63×10^{-1}
^{131}I hippuran	5.20×10^{-2}	0.74	3.85×10^{-2}
^{131}I MIBG	1.40×10^{-1}	0.74	1.0×10^{-1}
^{131}I sodium iodide (0% uptake)	6.10×10^{-2}	3700	n.a.
^{131}I sodium iodide (35% uptake)	2.40×10^{2}	3700	n.a.
^{111}In pentetreotide, also known as Octreoscan	5.40×10^{-2}	222	1.20×10^{1}
^{111}In white blood cells	3.60×10^{-1}	18.5	6.66×10^{0}
^{81m}Kr krypton gas	2.70×10^{-5}	370	9.99×10^{-3}
^{15}O water	9.30×10^{-4}	370	3.44×10^{-1}
^{32}P phosphate	2.40×10^{0}	148	3.55×10^{2}
^{153}Sm lexidronam, also known as quadramet	1.97×10^{-1}	2590	-
^{89}Sr chloride, also known as Metastron	3.10×10^{0}	148	-
^{99m}Tc apcitide, also known as AcuTect	9.30×10^{-3}	740	6.88×10^{0}
^{99m}Tc depreotide, also known as NeoTect	2.30×10^{-2}	740	1.70×10^{1}
^{99m}Tc disofenin, also known as HIDA (iminodiacetic acid)	1.70×10^{-2}	185	3.15×10^{0}

Table 11. (Continued)

Procedure	Effective dose per administrated activity [mSv/MBq]	Administrated activity [MBq]	E [mSv]
^{99m}Tc DMSA (dimercaptosuccinic acid), also known as Succimer	8.80×10^{-3}	185	1.63×10^{0}
^{99m}Tc exametazime, also known as Ceretec and HMPAO	9.30×10^{-3}	740	6.88×10^{0}
^{99m}Tc macroaggregated albumin (MAA)	1.10×10^{-2}	148	1.63×10^{0}
^{99m}Tc medronate, also known as Tc-99m Methyenedi-phosphonate (MDP)	5.70×10^{-3}	740	4.22×10^{0}
^{99m}Tc mertiatide, also known as MAG3 (normal renal function)	7.00×10^{-3}	740	5.18×10^{0}
^{99m}Tc mertiatide, also known as MAG3 (abnormal renal function)	6.10×10^{-3}	740	4.51×10^{0}
^{99m}Tc mertiatide, also known as MAG3 (acute unilateral renal blockage)	1.00×10^{-2}	740	7.40×10^{0}
^{99m}Tc Neurolite, also known as ECD and Bicisate	1.10×10^{-2}	740	8.14×10^{0}
^{99m}Tc pentetate, also known as Tc-99m DTPA	4.90×10^{-3}	370	1.81×10^{0}
^{99m}Tc pyrophosphate	5.70×10^{-3}	555	3.16×10^{0}
^{99m}Tc red blood cells	7.00×10^{-3}	740	5.18×10^{0}
^{99m}Tc sestamibi, also known as Cardiolite (rest)	9.00×10^{-3}	740	6.66×10^{0}
^{99m}Tc sestamibi, also known as Cardiolite (stress)	7.90×10^{-3}	740	5.85×10^{0}
^{99m}Tc sodium pertechnetate	1.30×10^{-2}	370	4.81×10^{0}
^{99m}Tc sulphur colloid	9.40×10^{-3}	296	2.78×10^{0}

Procedure	Effective dose per administrated activity [mSv/MBq]	Administrated activity [MBq]	E [mSv]
^{99m}Tc Technegas	1.50×10^{-2}	740	1.11×10^{1}
^{99m}Tc tetrofosmin, also known as Myoview (rest)	7.60×10^{-3}	740	5.62×10^{0}
^{99m}Tc tetrofosmin, also known as Myoview (stress)	7.00×10^{-3}	740	5.18×10^{0}
^{201}Tl thallous chloride (with contaminants)	1.60×10^{-1}	74	1.18×10^{1}
^{133}Xe xenon gas (rebreathing for 5 minutes)	8.00×10^{-4}	555	4.44×10^{-1}

HEALTH EFFECTS AND RISK MAGNITUDE

X-rays are forms of ionizing radiation that may interact with atoms and can cause ionization in cells. They may produce free radicals or direct effects that can damage DNA or cause cell death. Health effects of ionizing radiation are classified into two types: those that are visible and manifested within a relatively short time (called tissue reactions or formerly deterministic effects: skin erythema, hair loss, cataract, infertility, circulatory disease) and others which are only estimated and may take years or decades to manifest (called stochastic effects: cancer and genetic effects). Tissue reactions have thresholds, which are typically quite high (Table 12). There are a large number of reports of skin injuries among patients from fluoroscopic procedures in interventional radiology and cardiology, CT procedures and combination of multiple procedures [40].

The lens of the eye is one of the more radiosensitive tissues in the body [103]. Radiation-induced cataract has been demonstrated among workers involved with interventional procedures using x-rays [103, 104]. A number of studies suggest there may be a substantial risk of lens opacities in populations exposed to low doses of ionizing radiation. These include patients undergoing CT scans [105] and some other population categories as, radiologic technologists/radiographers [106] atomic bomb survivors [107] and those exposed in the Chernobyl accident [108]. Up until recently, cataract formation was considered a tissue reaction with a threshold for detectable opacities of 5 Sv for protracted exposures and 2 Sv for acute exposures [14, 103]. Based on recent epidemiological evidences, the ICRP has now suggested that there are some tissue reaction effects, particularly those with very late manifestation, where threshold doses are or might be lower than previously considered. For the lens of the eye, the threshold in absorbed dose is now considered to be 0.5 Gy [14].

Table 12. Thresholds for tissue reactions [40]

Tissue and effect	Threshold	
	Total dose in a single exposure (Gy)	Annual dose if the case of fractionated exposure (Gy/y)
Testes		
Temporal sterility	0.1	0.4
Permanent sterility	6.0	2.0
Ovaries		
Sterility	3.0	>0.2
Lens		
Cataract (visual impairment)	0.5	0.5 divided by years duration
Bone marrow		
Depression of hematopoiesis	0.5	>0.4
Heart or brain *Circulatory disease*	0.5	0.5 (total dose for fractionated exposure)

Tissue reactions occur in the application of ionizing radiation in radiation therapy, and in interventional procedures, particularly when fluoroscopically guided interventional procedures are complex and require longer fluoroscopy times or acquisition of numerous images.

Stochastic effects include cancer and genetic effects, but the scientific evidence for cancer in humans is stronger than for genetic effects. According to ICRP Publication 103 [14], detriment-adjusted nominal risk coefficient for stochastic effects for whole population after exposure to radiation at low dose rate is 5.5% per Sv for cancer and 0.2% per Sv for genetic effects. This gives a factor of about 27 more likelihood of carcinogenic effects than genetic effects.

Radiation doses to patients from diagnostic and interventional procedures vary greatly. Thus, for assessment of risk emerging from there procedures, one must one must determine the dose. It important to mention that cancer risk estimates are based on models of a nominal standard human and cannot be considered to be valid for a specific individual person. At present, it is assumed that stochastic risks have no threshold, e.g, linear no-threshold relationship of dose-effect is valid. This means that at any level of radiation exposure, the risk is assumed to remain, no matter how small it is. The probability of a stochastic effect attributable to the radiation increases with dose and is probably proportional to dose at low doses. At higher doses and dose rates, the probability often increases with dose more markedly than simple proportion, which happens also when ionizing radiation is used in medical procedures. With the current state of knowledge, carcinogenic effects are more likely for organ doses in excess of 100 mGy [14, 16]. Although a single radiological examination only leads to a small increase in the probability of cancer induction in a patient, in developed countries each member of the population undergoes, on average, one such examination each year [1]. Therefore, the cumulative risk increases accordingly. For example, a chest CT scan that yields about 8 mSv effective dose can deliver about 20 mGy dose to the breast; 5 CT scans will therefore deliver about 100 mGy. There may be controversies about cancer risk at the radiation dose from one or a few CT scans, but the doses encountered from 5 to 15 CT scans approach the exposure levels where risks have been documented [40]. Some groups of patients are examined much more frequently due to their health status. Also, some groups show higher than average sensitivity for cancer induction (e.g., embryo/fetus, infants, and young children, those with genetic susceptibility). Also, it is well known that different tissues and organs have different radiosensitivities. For example, females are more radiosensitive than males and young patients are more sensitive than older patients. For example, if a man 40 years old is exposed to radiation, his risk of lung cancer is estimated to be 17% higher than if he was exposed to the same radiation dose at age 60 [16, 40]. All these circumstances indicate that proper justification of radiation use and optimization of radiation protection in medicine are indispensable principles of radiological protection [16].

DOSE MANAGEMENT FOR HIGH DOSE PROCEDURES

While new, digital techniques have the potential to reduce patient doses, they also have the potential to significantly increase them. A balance between radiation dose and image quality is needed. Optimization does not mean simply maximizing image quality and minimizing patient dose; rather, it requires radiologists or other physician to determine the level of image quality that is necessary to make the clinical diagnosis and then for the dose to be minimized without compromising the image quality [5]. Diagnostic reference levels should be set up and refined, specific for clinical image quality and adjusted for body weight/size and clinical task.

As already discussed, different clinical tasks require different levels of image quality. The doses that have no additional benefit for the clinical purpose should be avoided. All of these new challenges should be part of the optimization process and should be included in clinical and technical protocols. Local diagnostic reference levels should be re-evaluated for digital imaging, and patient dose parameters should be displayed at the operator console. Frequent patient dose audits should occur when digital techniques are introduced. Training in the management of image quality and patient dose in digital radiology is necessary, in particular during the transition from analogue to digital radiology. In addition, as digital images are easier to obtain and transmit, the justification criteria should be reinforced. In Publication 93 [109], the ICRP deals with the dose management in digital radiology.

With the increased availability of and reliance on high-dose imaging modalities for rapid and comprehensive diagnosis and treatment, the need for better radiation dose management has increased. National, international regulatory authorities and research groups investigate ways for reducing patient exposure and optimizing scanning protocols [110-128]. Justification and optimisation as well as the development of reference dose values are particularly important, especially in high-dose procedures (interventional radiology/cardiology and CT) and among the most sensitive population groups (paediatrics) [113].

CT scanners have been subject to major technological advances. This has greatly extended the role of CT imaging, which is now regarded as the most preferable radiological investigation for a broad spectrum of clinical symptoms, such as CT angiography, CT of the urinary system and especially head CT scan have increased substantially. Also, the introduction of clinical procedures combining more than one section of the body at a time, for example, scanning the whole trunk in a chest–abdomen–pelvis scan has implications on population dose [113]. CT scanning is now frequently used for the diagnosis and management of patients with complex cardiovascular disease with a high level of detail [115], pulmonary embolism [116] and CT guided interventional procedures [117]. Spiral CT has also greatly improved the evaluation of children's diseases, and it is particularly useful for diagnosis and assessment of congenital heart abnormalities, lung and airway diseases, trauma and infections [118-120]. Therefore, the monitoring of trends in CT patient doses is currently particularly important, especially in paediatrics, as the organ doses delivered from a common CT scan result in an increased risk of radiation induced carcinogenesis, particularly for children [121].

CT is a valuable modality in which the benefits generally outweigh individual risks. However, as CT images are easy, painless and quick to obtain, and since doses are among the highest from all diagnostic procedures, the justification criteria should be reinforced and careful consideration of the use of CT is necessary, given the increasing frequency of such examinations. Many initiatives for dose reduction in paediatric CT has been taken worldwide. The International Atomic Energy Agency (IAEA) has performed an international radiation dose survey in paediatric CT that was undertaken as part of technical cooperation projects [110-111]. From the other side The Image Gently Alliance is a coalition of health care organizations dedicated to providing safe, high quality paediatric imaging worldwide. The primary objective of the Alliance is to raise awareness in the imaging community of the need to adjust radiation dose when imaging children. The ultimate goal of the Alliance is to change practice. One of the results of Image gently campaign is the article on how to optimize and adjust the CT protocols for children [122]. This article suggests 10 steps that radiologists and radiologic technologists, with the assistance of their medical physicist, can take to obtain good quality CT images while properly managing radiation dose for children undergoing CT. The first six steps ideally should be completed before performing any CT on a paediatric patient and the final four steps address the unique consideration that should be given for each scanned patient:

1. Increase Awareness and Understanding of CT Radiation Dose Issues Among Radiologic Technologists
2. Enlist the Services of a Qualified Medical Physicist
3. Obtain Accreditation
4. When Appropriate, Use an Alternative Imaging Strategy That Does Not Use Ionizing Radiation
5. Determine if the Ordered CT Is Justified by the Clinical Indication
6. Establish Baseline Radiation Dose for Adult-Sized Patients
7. Establish Radiation Doses for Paediatric Patients by "Child-Sizing" CT Scanning Parameters
8. Optimize Paediatric Examination Parameters
9. Scan Only the Indicated Area: Scan Once
10. Prepare a Child-Friendly and Expeditious CT Environment

Large variation in doses has been reported in adult and paediatric CT for a similar type of procedure, especially in paediatric examinations where the range of patient size is wider within the same age band [123, 124]. This arises not only from geometrical differences but also from differences in beam filtration, X-ray spectrum and beam profile but could be eliminated by applying an optimised protocol [125]. Dose variations in CT are mainly due to different scanner types/manufacture and different scanning protocols as

Radiation Exposure in Medical Imaging   51

well as due to variations in the selected length of the region to be scanned, tube rotation speed, helical pitch, collimation, filtration, patient weight, etc. [126].

As already described, the use of Diagnostic Reference Levels (DRLs) has been proposed as an optimisation tool, as it identifies high dose practices where dose-reduction techniques would have the greatest impact [124, 127]. DRLs are expressed in terms of Computed Tomography Dose Index (CTDI), dose-length product (DLP) and effective dose (E) for adults for each examination [75]. DRLs for paediatric CT examinations have been also proposed in terms of DLP, CTDI and E [17, 127].

For complex fluoroscopically guided interventional procedures, the management of procedures to conserve the application of radiation is essential to minimize the risk of injury. This concept takes on a wide variety of perspectives because it must be adapted to the wide variety of equipment and environments that might be employed for these procedures. Understanding the limitations of equipment while limiting radiation delivery from the specific equipment used in the procedure is an important goal to keep the risk reasonably low. Additionally, understanding the characteristics of the patient and of the procedure that can lead to injury is another essential factor in assisting the physician in radiation management [128]. Fluoroscopically guided interventional procedures are being increasingly used by a number of specialties as radiology, cardiology, vascular surgery, orthopaedic surgery, gastroenterology and many others. Due to complexity of these procedures, often associated with increased fluoroscopy time, some patients are suffering radiation induced skin injuries, while younger patients may face an increased risk of future cancer. Radiation induced skin injuries are occurring in patients due to the use of inappropriate equipment but more likely due to inadequate operational technique.

ICRP Publications 85 and 117 [40, 65] addresses the challenge of avoiding radiation injuries from medical interventional and fluoroscopy guided procedures. The absorbed dose to the patient in the area of skin that receives the maximum dose is of priority concern. Each local clinical protocol should include, for each type of interventional procedure, a statement on the cumulative skin doses and skin sites associated with the various parts of the procedure. While some x-ray might be equipped with dose monitoring devices, many are still not. To assist physicians in dose management, monitoring dose to the patient is essential. Understanding the advantages and limitations of dose measuring features as available in the machine is critical to dose optimization. Understanding how to monitor dose in the absence of integral devices is another important challenge. Finally, the training of the individual in the performance of a procedure is an essential factor in optimization. Experience contributes to the efficient completion of a procedure and is important in the optimization of the benefit-risk ratio [129]. Therefore, many researchers have reported methods used to quantify radiation delivery to patients in interventional procedures that can be classified as: (1) Real-time or post-procedure readout devices, (2) Skin surface or machine output devices, (3) Local small area monitors or wide-area monitors, (4) Direct or indirect monitors [66-68, 112].

Therefore, interventionists should be trained to use information on skin dose and on practical techniques to control dose. Maximum cumulative doses should be recorded in the patient record, and there should be a patient follow-up procedure for such cases. Risks and benefits, including radiation risks, should be taken into account when new interventional techniques are introduced. Careful selection of examination protocol is crucial, owing that all skin injuries can be avoided by a proper selection of radiographic technique. Some of the aspects that have to be taken into account are [17, 40]: positioning, collimation, selection of exposure factors, the protocol needs tailored to the patient size, fields need to be tightly aligned to the area of interest, the image intensifier and/or receptor needs to be positioned over the area of interest before fluoroscopy is commenced, field overlap in different runs and electronic magnification needs to be minimized, added copper filtration and pulsed fluoroscopy needs to be used, the number and timing of acquisitions, contrast parameters, patient positioning and suspension of respiration need to be planned, the patient table needs to be kept as far from the x-ray source as possible and the image intensifier and/or receptor needs to be as close to the patient as possible and fluoroscopy time needs to be limited.

CONCLUSION

Diagnostic and interventional procedures cover a diverse range of examination types, many of which are increasing in frequency and technical complexity. Due to the dominant contribution to population dose from manmade sources of radiation, it is important that radiation risks and benefits associated with medical exposure are well understood and that the basic principles of radiation protection for patients are implemented.

With respect to the trend in dosimetry and metrology in the field of ionizing radiation medical applications to calibrate dosimeters in the conditions that are similar to the clinical environment, routines for calibration in terms of air kerma, kerma-area product and kerma-length product for dosimeters used in conventional radiography, fluoroscopy, mammography and computed tomography are developed, with an emphasis on specific radiation qualities, calibration set up and uncertainty assessment. Objectives of clinical dose measurements in diagnostic and interventional radiology are multiple, as assessment of equipment performance, optimization of practice through the establishment of diagnostic reference levels or assessment of risk emerging from the use of ionizing radiation. Therefore, from the clinical point of view, the requirements for dosimeters and procedures to assess dose to standard dosimetry phantoms and patients in diverse modalities. Inevitably, the medical field as well as knowledge about radiation risk has improved over the last decade. Moreover, the technical development significantly improved the diagnostic and therapeutic possibilities. However, there are still opportunities for improvement. As the ultimate goal is to arrive at a situation where medical radiation protection is evidence based, there is a need to narrow the gap between evidence and practice. For this purpose, more emphasis has to be devoted to basic radiation protection principles as well as to the risk assessment, long term follow-up and risk management.

REFERENCES

[1] United Nations, *Sources and Effects of Ionizing Radiation* (2008 Report to the General Assembly, with Scientific Annexes A and B), Scientific Committee on the Effects of Atomic Radiation (UNSCEAR), UN, New York, 2010.

[2] National Council on Radiation Protection and Measurements. *Ionizing radiation exposure of the population of the United States*. NCRP report no. 160. Bethesda, MD: National Council on Radiation Protection and Measurements, 2009.

[3] Holmberg, O., Malone, J., Rehani, M., et al. *Eur J Radiol*. 2010, 76, 15-19.

[4] International Commission on Radiological Protection. *Managing Patient Dose in Multi-Detector Computed Tomography (MDCT)*, ICRP Publication 102, Ann. ICRP 37 (1), 2007.

[5] International Atomic Energy Agency. *Radiation protection in medicine: setting the scene for the next decade:* proceedings of an International Conference, Bonn, 3–7 December 2012. — Vienna, International Atomic Energy Agency, 2015.

[6] Rehani, M., Ciraj-Bjelac, O., Al-Naemi, HM., et al. *Eur J Radiol*. 2012, 81, e982-e989.

[7] Rehani, M, Ciraj-Bjelac, O. *CT dose perspectives and initiatives of the IAEA*, in Radiation Dose from Adult and Paediatric Multidetector Computed Tomography, 2nd ed, Editors D. Tack, M. Kalra, P.A. Gevenois, Springer, 2012, pp 495-509.

[8] Muhogora, W.E., Ahmed, N.A., Beganovic, A., Benider, A., Ciraj-Bjelac, O., et al., *Radiat. Prot. Dosim.* 2010, 140, 49-58.

[9] Ciraj-Bjelac, O., Beganovic, A., Faj, D., et al. *Eur J Radiol*. 2011, 79, e70-e73.

[10] Ciraj-Bjelac, O., Faj, D., Stimac, D., et al. *Eur J Radiol*. 2011, 78, 122-128.

[11] Jablanovic, D., Ciraj Bjelac, O., Damjanov, N, et al., *Radiat Prot Dosim*, 2013,155, 88-95.

Radiation Exposure in Medical Imaging 55

[12] International Commission on Radiological Protection. *1977 Recommendations of the ICRP*. ICRP Publication 26. Ann. ICRP 1 (3), 1977.

[13] International Commission on Radiological Protection. *1990 Recommendations of the International Commission on Radiological Protection*. ICRP Publication 60. Annals of the ICRP 21(1-3). Pergamon Press, Oxford, 1991.

[14] International Commission on Radiological Protection. *The 2007 Recommendations of the International Commission on Radiological Protection*. ICRP Publication 103, Ann. ICRP 37 (2-4), 2007.

[15] Wrixon, A. D. *J. Radiol. Prot.* 2008, 28, 161–168.

[16] International Commission on Radiological Protection. *Radiological Protection in Medicine*. ICRP Publication 105. Ann. ICRP 37 (6), 2007.

[17] International Atomic Energy Agency. *Radiation protection in paediatric radiology*. Vienna, 2012.

[18] European Commission, *Referral Guidelines for Imaging*, Radiation Protection 118, Office for Official Publications of the European Communities, Luxembourg, 2000.

[19] American College of Radiology, *ACR Appropriateness Criteria*, 2013, http://www.acr.org/Quality-Safety/Appropriateness-Criteria.

[20] Royal College Of Radiologists, iRefer: *Making the Best Use of Clinical Radiology*, 7th edn, The Royal College of Radiologists, London, 2012.

[21] Government of Western Australia, Department of Health, *Diagnostic Imaging Pathways*, 2013, http://www.imagingpathways.health.wa.gov.au/.

[22] International Atomic Energy Agency. *Comprehensive Clinical Audits of Diagnostic Radiology Practices: A Tool for Quality Improvement*, Vienna, 2010.

[23] International Atomic Energy Agency. *Radiological protection for medical exposure to ionizing radiation: safety guide jointly sponsored by the International Atomic Energy Agency*, the Pan American Health Organization and the World Health Organization, Vienna, 2002.

[24] Martin, C., Le Heron, J., Borras, C. et al., *J. Radiol. Prot.* 2013, 33,711–734.

[25] Ciraj, O., Kosutic, D., Kovacevic, M. et al. *Physica Medica.* 2005, 21, 159-163.

[26] Ciraj, O., Markovic, S., Kosutic, D. *Radiat Prot Dosim.* 2005, 113, 330-335.

[27] Ciraj-Bjelac, O., Beganović, A., Faj, D. *Radiat Prot Dosim.* 2011, 147 (1-2), 62-67.

[28] Ciraj-Bjelac, O., Beciric, S., Arandjic, D., et al. *Radiat Prot Dosim.* 2010, 140, 75-80.

[29] Ciraj-Bjelac, O., Avramova-Cholakova, S., Beganovic, A. et al. *Eur J Radiol.* 2011, 81, 2161-8.

[30] Ciraj, O., Markovic, S., Kosutic, D. *Radiat Prot Dosim.* 2005, 114, 158-163.

[31] Hadnadjev, D.R., Arandjić, D.D., Stojanović, S.S., et al. *Nuclear Technology and Radiation Protection.* 2012, 305-310.

[32] Institute of Physics and Engineering in Medicine DRL Working Party. *Guidance and Use of Diagnostic Reference Levels for Medical X-Ray Examinations.* IPEM Report 88, IPEM, York, 2004.

[33] Rehani, M. *Br J Radiol.* 2015, 88, 20140344.

[34] International Commission on Radiation Units And Measurements, *Patient Dosimetry for X Rays Used in Medical Imaging*, ICRU Rep. 74, ICRU, Bethesda, MD, 2006.

[35] International Commission on Radiation Units And Measurements, *Fundamental Quantities and Units for Ionizing Radiation*, ICRU Rep. 60, ICRU, Bethesda, MD, 1998.

[36] International Atomic Energy Agency. *Technical report series 457: Dosimetry in diagnostic radiology – an international code of practice*, Vienna, 2007.

[37] International Electrotechnical Commission, Medical Electrical Equipment Part 2-44: *Particular Requirements for the Safety of X-Ray Equipment for Computed Tomography*, Rep. IEC-60601-2-44, IEC, Geneva, 2002.

Radiation Exposure in Medical Imaging 57

[38] European Commission, *European Guidelines for Quality Criteria for Computed Tomography,* Rep. EUR 16262, EC, Luxembourg, 2000.

[39] International Commission on Radiation Units and Measurements, *Quantities and Units in Radiation Protection Dosimetry*, ICRU Rep. 51, ICRU, Bethesda, MD, 1993.

[40] Rehani, M., Ciraj-Bjelac, O., Vaño, E., et al. *Radiological Protection in Fluoroscopically Guided Procedures outside the Imaging Department*, ICRP Publication 117, Ann. ICRP 40(6), 2010.

[41] Megzifene, A. Dance, D., McLean, D. et al. *Eur J Radiol.* 2010, 76, 11-14.

[42] Pernicka, F., Andero, P., Meghzifene, A. et al. Standards for radiation protection and diagnostic radiology at the IAEA dosimetry laboratory. IAEA SSDL Newletter, No 41, 1999, 13-26.

[43] International Atomic Energy Agency. *SSDL Network Charter.* The IAEA/WHO network of SSDL, IAEA, Vienna, 1999.

[44] International Atomic Energy Agency. *Absorbed dose determination in external beam radiotherapy: an international code of practice for dosimetry based on standards of absorbed dose to water*, Technical Report Series No 398, IAEA, Vienna, 2000.

[45] International Atomic Energy Agency. *Calibrations of radiation protection monitor instruments.* Technical Report Series No 16, IAEA, Vienna, 1999.

[46] Zoetelief, J., Pernicka, F., Alm Carlsson, G., et al. *Dosimetry in diagnostic and interventional radiology: International Commission on Radiation Units and Measurements and IAEA activities. Standards and Codes of Practices in Medical Radiation Dosimetry.* Proceedings of an International Symposium 25-28 November 2002, Vienna, Austria Vol 1, Vienna, IAEA, 387-404

[47] International Electrotechnical Commission. *Medical electrical equipment-dosimeters with ionization chambers and/or semiconductor detectors as used in X-ray diagnostic imaging.* Geneva: IEC-61674; 2012.

58 Danijela Arandjic, Predrag Bozovic, Olivera Ciraj-Bjelac et al.

[48] International Electrotechnical Commission. *Medical diagnostic X ray equipment - Radiation conditions for use in the determination of characteristics*, IEC- 61267, IEC, Geneva, 2004.

[49] Bochud, F., Grecescu, M., Valley, J F. *Phys Med Biol.* 2001, 46, 2477-2487.

[50] Larsson, P. et al., *Phys Med Biol.* 1996, 41, 2381-98.

[51] Ciraj Bjelac, O., Petrovic, B., Todorovic, N., et al. *Application of gamma radiation in medicine, in Gamma Rays: Technology, Applications and Health Implications*, ed. Istvan Bikit, Nova Publishers, 2012, 13, 321-344.

[52] International Commission on Radiation Units and Measuremnst. *Particle Counting in Radioactivity Measurements.* ICRU report 52, ICRU, 1992.

[53] Shrimpton, P. C., Wall, B. F., Fisher, E. S. *Phys Med Biol.*, 1981. 26, 133.

[54] Jones, D. G., Wall, B. F. *Organ doses from medical x-ray examinations calculated using Monte Carlo techniques* (No. NRPB-R--186). National Radiological Protection Board, 1985.

[55] Petoussi-Henss, N., Zankl, M., Fill, U., et al. *Phys Med Biol.*, 2001, 47, 89.

[56] Hart, D., Jones, D. G., Wall, B. F. *Estimation of effective dose in diagnostic radiology from entrance surface dose and dose-area product measurements* (No. NRPB-R--262). National Radiological Protection Board, 1994.

[57] Yang, C. H., Wu, T. H., Lin, C. J. et al. *Eur J Radiol.* 2016, 85, 1757-1764.

[58] Rahanjam, H., Gharaati, H., Kardan, M., et al. *Caspian Journal of Health Research.* 2016, 2, 18-27.

[59] Rasuli, B., Mahmoud-Pashazadeh, A., Ghorbani, M. et al. *Journal of Applied Clinical Medical Physics.* 2016, 17, 374-386.

[60] Ben-Shlomo, A., Bartal, G., Mosseri, M. et al. *The Spine Journal.* 2016, 16, 558-563.

Radiation Exposure in Medical Imaging 59

[61] British Institute of Radiology. Radiation Protection in Interventional Radiology (K. Faulkner and D. Teunen, eds.). British Institute of Radiology, London, 1995.

[62] Waite, J. C., Fitzgerald, M. *Radiat Prot Dosim.* 2001, 94, 89-92.

[63] Carlsson, C. A., Carlsson, A. (1990). *Dosimetry in diagnostic radiology and computerized tomography.* The dosimetry of ionizing radiation, vol. III (Kase KR, Bjärngard BE, Attix FH, Eds). Computed Tomography - fundamentals, system technology, image quality, Orlando: Academic Press, Chapter 2, 163-257.

[64] Chapple, C. L., Broadhead, D. A., Faulkner, K. *Br J Radiol,* 1995, 68, 1083-1086.

[65] International Commission on Radiological Protection. *Avoidance of Radiation Injuries from Medical Interventional Procedures.* ICRP Publication 85. Ann. ICRP 30 (2), 2000.

[66] Farah, J., Trianni, A., Ciraj-Bjelac, O. et al. *Medical Physics.* 2015, 42, 4211-4226.

[67] Dabin, J., Negri, A., Farah, J., Ciraj-Bjelac, O. et al. *Phys Med.* 2015, 31, 1112-7.

[68] Farah, J., Trianni, A., Carinou, E., Ciraj-Bjelac, O. et al. *Radiat Prot Dosim.* 2015, 164,138-142.

[69] Tsapaki, V., Kottou, S., Fotos, N. et al. *OMICS J Radiol,* 2017, 6, 1-7.

[70] Kastrati, M., Langenbrink, L., Piatkowski, M., et al. *Am J Cardiol,* 2016, 118, 353-356.

[71] Andreou, K., Pantos, I., Tzanalaridou, E., et al. *Physica Medica.* 2016, 32, 234.

[72] Etard, C., Bigand, E., Salvat, C. et al. *European Radiology.* 2017, doi: 10.1007/s00330-017-4780-5.

[73] Antic, V., Ciraj-Bjelac, O., Rehani, M. et al. *Radiat Prot Dosim.* 2013, 154, 76-84.

[74] Mori, S., Nishizawa, K., Ohno, M., et al. *Br J Radiol.* 2006, 79, 888-92.

[75] Bongartz, G. G. S. J., Golding, S. J., Jurik, A. G., et al. European guidelines for multislice computed tomography. European Commission, 2004.

[76] Knipp, D., Lane, B. F., Mitchell, J. et al. *Journal of Computer Assisted Tomography*. 2017, 41, 41-147.

[77] Aliasgharzadeh, A., Mihandoost, E., Mohseni, M. *Journal of Cancer Therapeutics and Research.*2017, 41, 141-147.

[78] Willemink, M. J., Leiner, T., de Jong, P. A. et al. *European radiology*. 2013, 23, 1632-1642.

[79] Martini, K., Barth, B. K., Nguyen-Kim, T. D. et al. *European journal of radiology*. 2016, 85, 360-365.

[80] Ghetti, C., Ortenzia, O., Palleri, F. et al. *Radiat Prot Dosim*. 2016, DOI: 10.1093/rpd/ncw264.

[81] Mainiero, M. B., Lourenco, A., Mahoney, M. C. et al. *Am Coll Radiol*. 2016, 13, R45-R49.

[82] Chetlen, A. L., Brown, K. L., King, S. H. et al. *AJR Am J Roentgenol*. 2016, 206, 359-365.

[83] Widmann, G., Al-Shawaf, R., Schullian, P. et al.. *European Radiology*. 2017, 27, 2225-2234.

[84] Hatziioannou, K., Psarouli, E., Papanastassiou, E. et al. *Dentomaxillofacial Radiology*. 2014, 34, 304-307.

[85] Rehani, M. M., Gupta, R., Bartling, S., et al. ICRP Publication 129: Radiological Protection in Cone Beam Computed Tomography (CBCT), 2015.

[86] Strocchi, S., Colli, V., Conte, L. *Radiat Protection Dosim*. 2012, 151, 162-165.

[87] Brisco, J., Fuller, K., Lee, N. et al.. *British Journal of Oral and Maxillofacial Surgery*. 2014, 52, 76-80.

[88] Dierckx, D., Vargas, C. S., Rogge, F. et al. *Radiat Protection Dosim*, 2015, 163, 125-132.

[89] Kyriakou, Y., Richter, G., Dörfler, A. et al. *American Journal of Neuroradiology*. 2018, 29, 1930-1936.

[90] Petersen, A. G., Eiskjær, S., Kaspersen, J. *Pediatric radiology*. 2012, 42, 965-973.

[91] Wielandts, J. Y., Smans, K., Ector, J. et al. *Phys Med Biol*. 2010, 55, 563-79.

Radiation Exposure in Medical Imaging 61

[92] Granlund, C., Thilander-Klang, A., Ylhan, B., et al. *Br J of Radiol.* 2016, 89, 20151052.

[93] Scarfe, W. C., Azevedo, B., Toghyani, S. et al. *Australian dental journal.* 2017, 62, 33-50.

[94] International Commission on Radiological Protection. *Radiation Dose to Patients from Radiopharmaceuticals.* ICRP Publication 106, 2007.

[95] International Commission on Radiological Protection. *Radiation Dose to Patients from Radiopharmaceuticals.* ICRP Publication 80. Annals of the ICRP 28(3). Pergamon Press, Oxford, 1998.

[96] International Commission on Radiological Protection. *Radiation Dose to Patients from Radiopharmaceuticals.* ICRP Publication 53. Annals of the ICRP 18(1-4). Pergamon Press, Oxford, 1988.

[97] Tolboom, N., Berendse, H. W., Leysen, J. E. et al. *Neuropsychopharmacology.* 2015, 40, 472-479.

[98] Aalto, S., Ingman, K., Alakurtti, K. et al. *Journal of Cerebral Blood Flow & Metabolism.* 2015, 35, 424-431.

[99] Jansen, N. L., Suchorska, B., Wenter, V. et al., *Journal of Nuclear Medicine.* 2015, 56, 9-15.

[100] Everitt, S. J., Ball, D. L., Hicks, R. J. et al., *Journal of Nuclear Medicine.* 2014, 55, 1069-1074.

[101] Calabria, F., Chiaravalloti, A., Schillaci, O., *Clinical nuclear medicine.* 2014, 39, 122-130.

[102] Sanchez-Crespo, A., Christiansson, F., Thur, C. K. et al. *European Journal of Nuclear Medicine and Molecular Imaging.* 2016, 1-8.

[103] International Commission on Radiological Protection, *ICRP Statement on Tissue Reactions / Early and Late Effects of Radiation in Normal Tissues and Organs – Threshold Doses for Tissue Reactions in a Radiation Protection Context.* ICRP Publication 118. Ann. ICRP 41(1/2), 2012.

[104] Ciraj-Bjelac, O., Rehani, M.M., Sim, K.H., et al., *Catheter. Cardiovasc. Interv.* 2010, 76, 826-834.

[105] Klein, B.E., Klein, R., Linton, K.L., et al., *Am J Public Health.* 1993, 83, 588-590.

62 Danijela Arandjic, Predrag Bozovic, Olivera Ciraj-Bjelac et al.

[106] Chodick, G., Bekiroglu, N., Hauptmann, M., et al. *Am J Epidemiol.* 2008, 168, 620-631.

[107] Nakashima, E., Neriishi, K., Minamoto, A., et al. *Health Phys.* 2006, 90, 154-160.

[108] Day, R., Gorin, M.B., Eller, A.W. Health Phys. 1995, 68, 632-642.

[109] International Commission on Radiological Protection, *Managing Patient Dose in Digital Radiology.* ICRP Publication 93, Ann. ICRP 34 (1), 2004.

[110] Vassileva J. et al. *AJR The Am J Roentgenol.* 2012, 198, 1021-1031.

[111] Vassileva J. et al. *Radiat Prot Dosim.* 2015, 165, 70-80.

[112] Padovani, R. et al. *Rad. Prot. Dosim. 2005,* 117, 217-221.

[113] Dougenia, E., Faulkner, K.,Panayiotakis, G. E. *Eur J Radiol,* 2012, 81, 665-683.

[114] American Association of Physicists in Medicine. *The measurement, reporting, and management of radiation dose in CT.* AAPM Task Group 23, Report 96. College Park, MD: AAPM; 2008.

[115] Kalender, W.A., Ulzheimer, S., Kachelrieß, M. *Cardiac imaging with multislice spiral computed tomography.* In: Scan reconstruction principles and quality assurance. Multislice CT: a practical guide. Heidelberg: Springer; 2001.

[116] Coche E, Vynckier S, Octave-Prignot M. *Radiology.* 2006, 240, 690–7.

[117] Tsalafoutas IA, et al. *AJR Am J Roentgenol.* 2007, 188, 1479–84.

[118] Frush, D.P. Donnelly, L.F., Rosen, N.S. *Pediatrics.* 2003,112, 951–7.

[119] Frush, D.P. *Pediatr Radiol.* 2008, 38, S259–66.

[120] Frush, D.P., Donnelly, L.*F. Radiology.* 1998, 209,37–48.

[121] Brenner, D., Elliston, C., Hall, E., Berdon, W.. *AJR Am J Roentgenol.* 2001, 176, 289–96.

[122] Strauss, K. et al. *AJR Am J Roentgenol.* 2010, 194, 868-873

[123] Verdun, F.R. et al. *Eur Radiol. 2008, 18, 1980–6.*

[124] Shrimpton, P.C, Wall BF. *Radiat Prot Dosim.* 2000, 90, 249–52.

[125] Shrimpton, P. C., Hillier, M.C., Lewis, M.A., et al. *Br J Radiol.* 2006, 79, 968–80.

[126] Hamberg, L.M., Rhea, J.T., Hunter, G.J. et al. *Radiology.* 2003, 762–72.

[127] McNitt-Gray, M.F. *Radiographics.* 2002, 22, 1541–53.

[128] Kubo T, Lin PJ, Stiller W, et al. Radiation dose reduction in chest CT: a review. *AJR Am J Roentgenol* 2008, 190(2), 335–43.

[129] International Atomic Energy Agency. Patient Dose Optimization in Fluoroscopically Guided Interventional Procedures. *IAEA-TECDOC-1641,* Vienna, 2010.

In: Radioactive Wastes and Exposure
Editor: Austin D. Russell

ISBN: 978-1-53612-213-8
© 2017 Nova Science Publishers, Inc.

Chapter 2

SOLID RADIOACTIVE WASTES IN NUCLEAR MEDICINE

Gabriela Hoff[1], PhD and Cláudia R. Brambilla[2]

[1]Dipartimento di Fisica, Università di Cagliari, Monserrato, Italy
[2]Institute of Neuroscience and Medicine, Medical Imaging Physics, (INM-4), Forschungszentrum Juelich GmbH, Jülich, Germany

ABSTRACT

This is a critical review about solid radioactive wastes in Nuclear Medicine facilities and its management, storage and disposal. In the last 10 years, the number of nuclear medicine facilities is growing up and the radioactive wastes need to be managed in a correct way to prevent environmental implications in their disposal. Storage of radioactive waste in medical facilities is a procedure to deal with relatively short physical half-life radionuclides (such as 8 days, 6 hours or 110 minutes to Iodine 131, Technetium 99m and Fluorine 18 respectively). This approach depends on many factors, such as segregation, type of radionuclide, mass, dose rate and residual activity, time of storage and date of disposal. Radioactive solid wastes are present in different kinds of medical materials in a Nuclear Medicine facility and are generally classified as low-level radioactive waste. Many countries have defined previously clearance levels for radionuclides and some of them are referred to statistically significant differences from background activity. It is the responsibility of the regulatory authority of each country to define clearance levels and site-specific discharge authorizations. Different directives around the world contain a list of nuclides with values of quantities (Bq) and concentrations of activity per unit mass (Bq/g) that should not to be exceeded in radioactive waste disposal. The practice in some countries for this procedure of estimating this quantity/activity may be based on theoretical calculations and estimating by using Geiger Müller measurement of exposure rate or the total storage time is based in 10% of the initial activity. This measure estimation using GM could be affected by many factors as energy dependence of the detector, radionuclide involved - energy spectra, geometry of the solid waste and it heterogeneity of the waste container or box, mixture of radionuclides in the segregation process, among others factors. In this critical review is presented the international regulations and procedures for management of solid radioactive waste generated in Nuclear Medicine and its evolution in time according as well as the current practice in different countries. It will presented results of Monte Carlo simulations and deterministic calculation to perform an evaluation of different factors that could affect the determination of activity concentration in the management of solid radioactivity wastes and its implications to discharge it correctly based on geometry of the solid radioactive waste container/box when is used Geiger Müller detectors.

Keywords: solid radioactive wastes, nuclear medicine, Geiger Müller, management

INTRODUCTION

Radioactive waste is produced in nuclear power plants and the use of radioactive materials in industry, agriculture, medicine, research and education. It is important a safe management program of radioactive waste to protect the human health and the environment.

The International Atomic Energy Agency (IAEA) have been establishing a program to set the principles and standards for the safe management of waste and publishing an important number of guidelines to support the safety management of the radioactive waste since 1958 (IAEA, 1958). The program aim is to set the documentation that reflects an international consensus. Each member state needs to have a national framework that sets the appropriate requirements for radioactive waste management. Because the variation of nature and volume of radioactive waste from different sources, the requirements for its safe management also vary among countries. Safe radioactive waste management relies on the application of technology and resources in a regulated manner, in accordance with international regulations to achieve the objective that exposure to ionizing radiation is defined for worker members and humans from the public. The environment must be also protected.

REVIEW ABOUT THE INTERNATIONAL REGULATION

The first criteria of classification assumed by IAEA in the Safety Series No. 1 was about the physical state of the waste, e.g., solid or liquid and depending of the installation necessities including the criteria about gamma radiation level (high, low), total activity (high, intermediate, low), half-life (short, long) and combustible or not combustible (IAEA, 1958). Gaseous state or aerosol was already mentioned and the regulation suggested filters to be treated as solid waste when the limits of disposable were not yet archived. Moreover, it is recommended record the quantities of radioactive wastes released to drainage systems, to sewers, or for burial. The storage classification was defined as temporary or indefinite and storage conditions were required according to the safety limit is provided by keeping the level of activity at the point of release into the environment below the permissible levels for no occupationally exposed persons recommended by the International Commission on Radiological Protection (ICRP) for activity in drinking water or in air in 1954 and 1956. The Table 1 shows some values from radionuclides that are still used in Nuclear Medicine (for diagnosis and treatment) and Positron Emission Tomography (PET) and the limits established in this first documentation.

Gabriela Hoff and Cláudia R. Brambilla

Table 1. Maximum Permissible Concentrations in Air and Water for Continuous Exposure defined by ICRP in 1954 and 1958

Radioisotope	Critical Organ	Maximum Permissible Concentration	
		In Water ($\mu c/cc.$)[a]	In Air ($\mu c/cc.$)[a]
^{18}F	Bone	0.2	3.10^{-5}
^{131}I	Thyroid	6.10^{-5}	6.10^{-9}
^{177}Lu	Bone	1.10^{-3}	2.10^{-7}
^{201}Tl	Muscle	9.10^{-3}	2.10^{-6}
^{133}Xe	Total Body	4.10^{-3}	4.10^{-6}

[a] Obsolete unit to express Curie per cubic centimeter
[1] Table adapted from IAEA Safety Series No.1 Appendix 1

It is general recommended by this first IAEA Safety Series documentation that disposal of radioactive wastes to the environment should be made in accordance with the conditions established by the "radiological health and safety officer" and by the authority in each country (IAEA, 1958). Moreover, the ways in which radioactive materials may affect the environment should be carefully examined until disposal methods being implemented. All the documentation published by IAEA Safety Standards is complemented by appropriate national regulatory infrastructures to be effective in each country. The IAEA produces a wide range of technical publications to help States in developing their national standards and infrastructures (IAEA, 2005). The IAEA publications are classified in different groups of reports in time, depending on the evolution and specificity the regulation and the knowledge in the field of interest. The first group of recommendation were the Safety Series – SS -, developed in the period of 1958 to 1997. The IAEA SS include principles, codes of practice, regulations, guidance, manuals and reports from panels of experts. From 1978 to 1990, the IAEA recommendations/publications were classified in four different SS categories: Safety Standards - SSS having basic safety standards, regulations and codes of practice; Safety Guides – SGS - supplementing information to SSS with recommendation of procedures to be followed; Recommendations on safety practices; Procedures and Data with information on procedures, techniques and criteria of interest on safety matters. Since 1990, the IAEA redefined the categories of publication being issued keeping the SSS and SGS documents and adding the: Safety Fundamentals - SFS - with basic objectives, concepts and principles to ensure safety; Safety Practices – SPS – with practical examples and detailed methods; Safety Recommendations – SRS – that are independent reports from expert groups; and Procedures and Data – SPDS - containing information on procedures, techniques and criteria pertaining to safety matters. In this chapter are presented some of the IAEA regulations, including the Recommendation (GSR) and IAEA technical documents (TECDOC). In Figure 1 and Figure 2 are presented summaries of the evolution of the IAEA regulations, but some interim and intermediate recommendation, about this subject, are not cited in the present text to avoid repetition of information and present a long and tedious text. It is important to note that there are other international and national regulations interesting and pertinent on radioactive waste disposal and management. For example, the IAEA comment and used, several times, the recommendations, concepts and limits defined previously by International Commission on Radiological Protection (ICRP) and United Nations Scientific Committee on the Effects of Atomic Radiation (UNSCEAR). The decision of presenting just IAEA recommendations was made to avoid prolixism and to keep following one philosophy of radiological protection applied to radioactive waste and show how the knowledge and concepts about this theme is evolving in time. To better understand the evolution in concepts and limits it will be presented, in Figure 1 and Figure 2, the summary scheme of IAEA regulations related to waste management (applied to medical facilities). Figure 1 shows the scheme of IAEA regulation in chronological organization that were already suppressed and its principal information (concept, methodology or limits) added or changed applied to medical facilities. Follow are commented the specific information about the changes on concepts and limits that can be applied to radioactive waste or can influence it management.

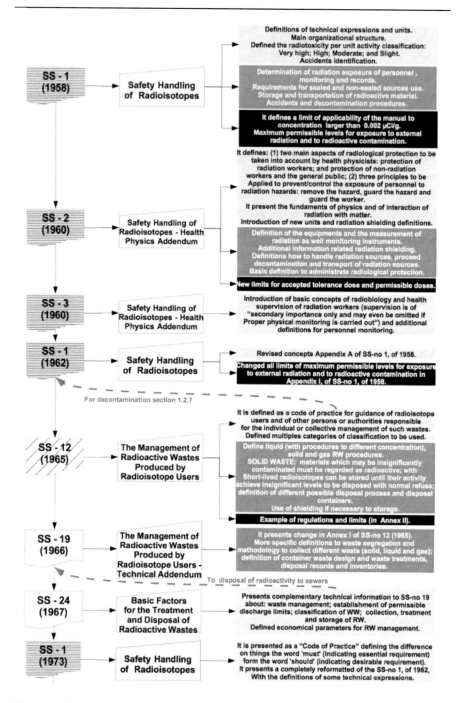

Figure 1. (Continued)

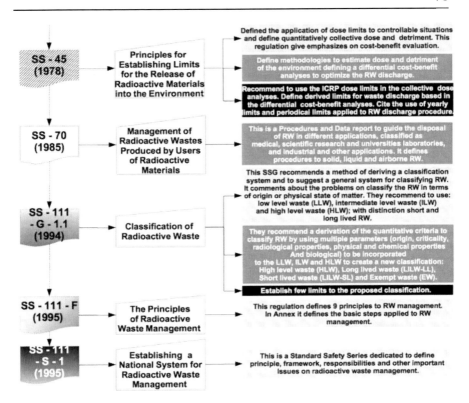

Figure 1. Chronological time line scheme of obsolete IAEA regulations. This scheme shows, in the three columns, the document identification, its name and the summary of the information document (light gray box refers to concepts, dark gray box to methods and black box to limits reported). The document identification that are filled represent the one that were substituted by a current regulation, presented in Figure 2, with the same filled area in the document identification.

The IAEA SS-1 (IAEA, 1958) was a guide to the safe handling of radioisotopes that was developed to all radioisotopes users, but was thought to help the small scale users who may not had direct access to other sources of information. This document defined a low limit degree of "radioactivity" (term used in the manual) of applicability of this recommendation as a concentration of 74 MBq/g (2 µCi/g) of material, or a total activity in the working area less than 3.7 kBq (0.1 µCi). Bellow these limits the so called dangerous radioisotopes are not the most dangerous. The IAEA SS-1 (IAEA, 1958) presented a classification of radioisotopes per radiotoxicity by unit of activity, listing the radioisotopes in Table 1 of this document (IAEA, 1958

– page 34) such as: very high, high, moderate, and slight. It had a specific section to radioactive dispose and control. In this section it is given direction of how to collect and deposit the solid and liquid radioactive waste, requesting a classification on its identification following the criteria: gamma radiation levels (high, low), total activity (high, intermediate, low), half-life (long, short), and combustible, non-combustible. About the disposal of radioactive waste it is defined that it must be done in accordance with the conditions established by the competent authority and when this authority does not provide any information the restrictive safe limit a suggestion is presented in Appendix I (IAEA, 1958 - pages 86 to 91) providing the level of activity at the point of release into the environment below the permissible levels for non-occupationally exposed persons recommended by the ICRP for activity in drinking water or in air (see examples of these limits in Table 1 previously presented in this Chapter). The IAEA SS-1 revised (IAEA, 1962 and IAEA, 1973) maintain the same concepts, classification, methodologies and limits defined in the previous version related to radioactive waste.

The SS-2 (IAEA, 1960) it was a specific recommendation applied to health physics (or medical physics). It presented a chapter about storage and disposal of radioactive waste. It defined the specific areas in a medical facility, specifying controlled areas to storage of radioactive waste and defined methodologies to dispose radioactive waste in the environment for different physics state: liquid (dilution, releasing of the liquid in small portions in long periods or releasing the wastes into natural waters having precaution to avoid general public directly or indirectly to contamination), and solid (by reducing its bulk, burial or temporary storage). It did not suggested limits for disposal, but requested that the activity must be reduced to those levels at which the waste may be disposed by conventional means and added that the release of radioactive waste should ensure compliance with existing national or international regulations.

The SS-3 (IAEA, 1960) did not have specific recommendations about radioactive waste, but in case of accidents defined that some products of accident and decontamination should be treated as radioactive waste.

The SS-12 (IAEA, 1965) was one recommendation specifically dedicated to deal with radioactive waste. The SS-12 reported several times to the previously technical report technology of radioactive waste management avoiding environmental disposal – TRS 27 (IAEA, 1964). The SS-12 (IAEA, 1965) had recommendations about the control, use of radioisotopes and the radioactive waste management and disposal, reporting some procedures for treatment, storage and environment containment of radioactive waste. In Annex I of this recommendation (IAEA, 1965 – pages 47 to 50) it is presented a list with types of waste associated with uses to different radioisotopes and no specific limit to dispose the radioactive waste was defined, but several important information and procedures, with examples of regulations of different countries, are presented in this recommendation.

The SS-19 (IAEA, 1966) was a technical addendum starting with the presentation of the table about the types of waste associated with some uses of a number of radioisotopes. It detailed the segregation and treatment methods and the collection of solid and liquid collection, the direct disposal of radioactive waste to sewer, and the airborne waste management. It presented some limits for incinerated solid radioactive waste with definition of specific concentration of activities in ashes and information about compression and burial solid radioactive waste without specify limits to those cases. This regulation presented detailed information about release radioactive waste in sewer (adapting the previous limits to a realistic condition), giving options of methodologies for liquid radioactive waste in medical area, considering the use of Iodine 131 and Phosphorus 32 to cancer treatment, to quantities for disposal that exceed the permissible values defined in Table III of this recommendation (IAEA, 1966 – page 32).

The SS-24 (IAEA, 1967) was focused on to define the radioactive waste management system, not discussing about limits or classification but bringing information about the essence of the management system. The SS-45 (IAEA, 1978) was applied to the radioactive waste produced by the fuel cycle of the nuclear power industry and products of nuclear activities. It defined the differential methodology to establish limits to discharge the radioactive waste. This recommendation was not applied to medical areas, because of the classification of the radioactive waste produced in this application, and it is not presented in Figure 1.

The SS-70 (IAEA, 1985) it was a Procedures and Data recommendation applied to radioactive waste for applications in hospitals, research laboratories, industry and agriculture. This recommendation proposed local (applied by the users) and national controls it presented a classification with the principal radionuclides used in medicine, clinical measurements and biological research with typical quantities per application, radioactive waste characterization and the mode of treatment of disposal. It had defined the management of solid, liquid and airborne radioactive waste too.

The SS-111-F (IAEA, 1995) presented radioactive waste management principles applied to radioactive material, as defined to be radioactive waste by the appropriate national authorities, and to the facilities used for the management of this waste from generation through disposal. It was a conceptual recommendation and defined the 9 principles applied to radiation waste management: protection of the human health, protection of the environment, protection beyond national borders, protection of future generations, burdens on future generations, national legal framework, control of radioactive waste generation, radioactive waste generation and management interdependencies, and safety of facilities.

In SS-111-G-1.1 (IAEA, 1994), a Safety Guide, is addressed to classify gaseous, liquid and solid radioactive waste, and identifies important characteristics associated to its management, considering the classification of a wide range of radioactive wastes:

> "from high level waste including spent nuclear fuel when it is considered radioactive waste, to wastes having such low levels of radioactivity that they cannot be considered as 'radioactive' and consequently can be safely disposed of without further nuclear regulatory control". (IAEA, 1994. p. 2)

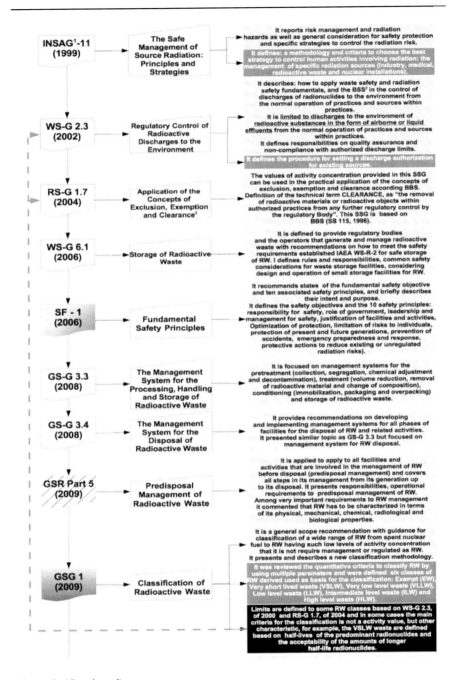

Figure 2. (Continued)

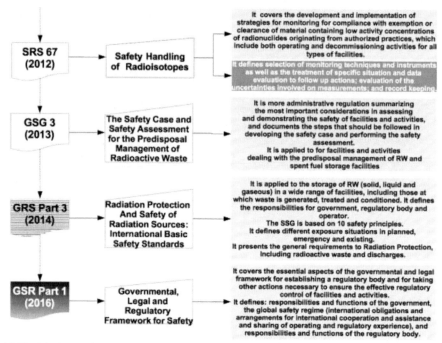

[1] INSAG is the anachronism to International Nuclear Safety Advisory Group.
[2] BBS is the anachronism to International Basic Safety Standards for Protection against Ionizing Radiation and for the Safety of Radiation Sources (BSS)
[3] This regulation is not related directly to the subject of interest, but the clearance concept introduced in this regulation is used nowadays.

Figure 2. Chronological time line scheme of the current IAEA recommendations in 2016. The three visible columns represents the identification of the document, its name and the summary of the information in the document (light gray box refers to concepts, dark gray box to methods and the black box to limits reported). The document identification that is filled represent the current substitute regulation to the obsolete ones presented in Figure 1.

The SS-111-G-1.1 (IAEA, 1994) proposed of a new classification methodology applied to radioactive waste with the following classes: high level waste (HLW); low and intermediate level waste (LILW) divided in two subgroups: the long lived waste (LILW-LL) and the short lived waste (LILW-SL); and exempt waste (EW). The characteristic of this classification are going to be presented following in the text to compare to the current classification.

The SS 111-S-1 (IAEA, 1995), a Safety Standard defined forth the elements to establish a national system to provide safe management for radioactive waste according national authorities. It did not set or review classification of radioactive waste neither established new limits, however defined the main objective for radioactive waste management and the principles on which radioactive waste management policy and strategies should be based; presented the basic components of a national framework for radioactive waste management; establish responsibilities of the Member State, the regulatory body and the waste generators and operators of radioactive waste management facilities; and described important features of radioactive waste management.

The Figure 2 is showing the summary scheme of the current version of IAEA regulation in 2016 related to waste management applied to medical facilities.

The management recommendation presented in Figure 2 (INSAG-11, 1999, GS-G-3.3, 2008; GS-G-3.4, 2008; GSR Part 5, 2009; GSG 3, 2014, GSR Part 3, 2014, and GRS Part 1, 2016) are dedicated to help the local and national authorities to create a management system adequate to the use, storage and dispose of radioactive waste. It is important to define concepts and principles for radioactive waste management and to create registration systems and methodologies to improve traceability of information about radioactive waste and its control, maintaining the liberty of the national government in define its own regulation and limits but keeping the safety of occupational workers, public in general and the environment.

The WS-G 2.3 and the RS-G 1.7 (IAEA, 2002 and IAEA, 2004) are recommendations that define concepts, classification and limits applied to radioactive waste disposal.

It is interesting to evaluate the criteria for radioactive waste classification to understand the organic-dynamicity of the concepts applied to radioactive waste management and the radiological protection in general. The complexity of the concepts can be observed when comparing the classifications presents in three different IAEA recommendations:

- *SS no. 1 (SS-1, 1958):* suggests classify radioactive waste according to methods of storage or disposal the facility may identify the radioactive waste by informing one or more of the following bases: (a) Gamma radiation levels (high, low); (b) Total activity (high, medium, low); (c) Half-life (long, short); and (d) Combustible or non-combustible.

- *SS-111-G-1.1 (IAEA, 1994):* suggests a classification based on multiples parameters: Exempt waste (EW) characterized by activity levels at or below clearance levels given in (SS no 89, 1988)[1] based on annual dose to members of the public less than 0.01 mSv; Low and intermediate level waste (LILW) presenting activity levels above clearance levels given in (IAEA, 1988)[1] and thermal power below about 2 kW/m^3, and it can be divided in two subgroups: the Long lived waste (LILW-LL) characterized by long lived radionuclide concentrations exceeding limitations for short lived waste and the Short lived waste (LILW-SL) restricted to long lived radionuclide concentrations (limitation of long lived alpha emitting radionuclides to 4000 Bq/g in individual waste packages and to an overall average of 400 Bq/g per waste package; and High level waste (HLW) that were characterized as waste of thermal power above about 2kW/m and long-lived radionuclide concentrations exceeding limitations for short lived waste.

- *GSG 1 (IAEA, 2009):* improved the classification based on multiple parameters and presents six levels (summarized in the scheme of Figure 3): exempt (EW) defined as radioactive waste that meets the criteria for clearance, exemption or exclusion from regulatory control for radiation protection purposes as described in RS-G-1.7 (IAEA, 2004)[2]; very short lived (VSLW) that includes radioactive

---

[1] In SS-89 of 1988 the exempt limits defined were: 10 µSv/year to individual dose per exempt practice; and 1 man.Sv/year for collective dose. Sources or practices resulting in doses below the exempt limits are considered inherently safe.

[2] In RSG-1.7 of 2004 the exempt limits defined were: effective doses for individuals in the order of 10 µSv or less in a year with an additional criterion named effective doses due to such low probability events should not exceed 1 mSv in a year; the collective effective dose commitments from one year of the practice will usually be well below 1 man.Sv. It is in agreement to the SS 115 definitions (IAEA, 1996).

waste containing primarily radionuclides with very short half-lives used in research and medical activities being limited to radioactive waste that can be stored for decay up to a few years and subsequently cleared for uncontrolled disposal, use or discharge; very low level (VLLW) considering the radioactive waste that does not meet the criteria of EW and does not need a high level of containment and isolation, being suitable for disposal in near surface landfill type facilities with limited regulatory control; low level (LLW) characterized by radioactive waste presenting values above clearance levels, but with limited amounts of long lived radionuclides, requiring robust isolation and containment for long period of time (few hundred years) and is suitable for disposal in engineered near surface facilities; intermediate level (ILW) represents the radioactive waste with long lived radionuclides that requires a greater degree of containment and isolation than that provided by near surface disposal; and High level (HLW) defined by radioactive waste with high levels of activity concentration capable to generate significant quantities of heat because of the radioactive decay process or composed by radioactive waste with large amounts of long lived radionuclides requiring a specific design of a disposal facility.

It is easy to observe, with the three examples of classification applied to radioactive waste, the evolution in time starting with a simple characterization of isolated parameters to and complex evaluation of multiples parameters that involve physical and chemical characteristics and considers the activity used in the facility such as some particularities of the facility activities developed.

Another concept largely used nowadays around the world (Nordic Council of Ministers, 1994; Menon, 2000 and Kosako 2000) in radioactive waste management is "Clearance". According to IAEA WS-G-2.3 (IAEA, 2000) clearance is based on the principle that some sources may be released from regulatory requirements since it can be demonstrated that they present trivial risks to individuals and populations. The Radiation Protection 122

(EURATOM, 2000) define that "Clearance" is thus reserved for release of material which does not require further regulatory control to ensure the actual destination of the material. The notion of specific clearance levels is introduced in this report for specific conditions which can be verified prior to release. In IAEA documents, including technical documents applied to specific situations (IAEA, 1996 and IAEA, 1998) the term "Clearance" is defined as the removal of radioactive materials or radioactive objects within authorized practices from any further regulatory control by the regulatory body (IAEA, 2004). As with exemption, the clearance may be granted by the regulatory body for release material produced by a specific practice. The radioactive materials that have no possibility of use are called as radioactive wastes and their legal position change from utilization system to waste management system. On very low-level radioactive wastes it is possible to ignore their human risk for their smallness of radiation impacts, and to use a clearance concept is reasonable to get an optimized use of radiation (IAEA, 2004). This technical term starts to be more often used in the mid of 80s based on a review of the IAEA Safety Series publications. In the mid of 90s, a major overhaul of the IAEA Safety Standards Program was initiated, with a revised oversight committee structure and a systematic approach to updating the entire corpus of standards (IAEA, 2009).

As one may observe the evolution of IAEA regulations presented several important issues related to radioactive waste management and safety, however one may extract two main issues related to radioactive waste and safety management: classification and clearance limits (or some quantitative limit to dispose). The purpose of classify the radioactive waste is simplify and help to plan the management, while the purpose of define clearance limits is to ensure safety. Clearance levels have to be based on scenarios encompassing exposure situations according to the relevant kind of material, quantities and uses for the material to be cleared. Moreover, the effort in developing precise limits has to be applied within the regulatory framework of licensing or authorizing specific radioactive waste management activities. Actual quantity or concentration limits for the classification of radioactive waste are to be established by the regulatory body of a member state. The standards about waste classification were also published by IAEA in 1970, and first reviews in 1981 and 1994. The standards published in 1970 and 1981 did not differ as much as the first comment about this in 1958 (already commented above). The standard from 1994 were published to cover some limitations in particular, the classification scheme lacked a completely coherent linkage to safety aspects in radioactive waste management, especially disposal. Three major aspects were included in the waste classification in this publication (IAEA, 2009):

1. Waste with so low concentration of radionuclides that it does not required regulatory control, and the radiological hazard is negligible (exemption or exclusion criteria).
2. Waste that contains such an amount of radioactive material that it requires actions to ensure the protection of workers and the public, either for short periods or for long periods of time.
3. Waste that contains such high levels of radioactive material that a high degree of isolation from the biosphere, usually by means of geological disposal, is required for long periods of time.

In 2009 the IAEA publication defined a six level scheme based on solid radioactive waste as demonstrated in the Figure 3.

In terms of safety, a radionuclide with a half-life less than 30 years is considered to be short lived (IAEA, 2009). It is important to make a distinction between waste containing short lived radionuclides and long lived radionuclides based on the radiological hazard associated. For example, in the short lived radionuclides the radiological hazard is significantly reduced by radioactive decay. Storage for decay is important for the clearance of radioactive waste. The practical experience shows that storage for decay is suitable for waste contaminated by radionuclides with a half-life of less than about 100 days (IAEA, 2006). The activity concentration of the waste should be determined and it is designated for decay after the segregation process (based on it classification), from the point of generation/raw waste up to the end of the decay storage defined period and its disposal. Usually, it is recommended to measure (e.g., samples) and analyzed prior to the removal of each batch from control based on regulatory body clearance definitions.

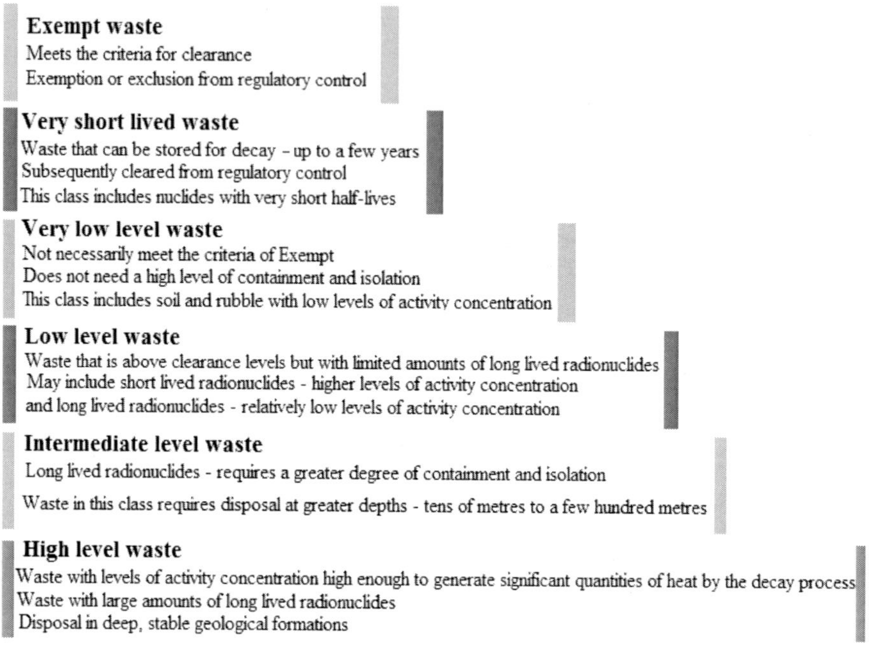

Exempt waste
Meets the criteria for clearance
Exemption or exclusion from regulatory control

Very short lived waste
Waste that can be stored for decay – up to a few years
Subsequently cleared from regulatory control
This class includes nuclides with very short half-lives

Very low level waste
Not necessarily meet the criteria of Exempt
Does not need a high level of containment and isolation
This class includes soil and rubble with low levels of activity concentration

Low level waste
Waste that is above clearance levels but with limited amounts of long lived radionuclides
May include short lived radionuclides - higher levels of activity concentration
and long lived radionuclides - relatively low levels of activity concentration

Intermediate level waste
Long lived radionuclides - requires a greater degree of containment and isolation
Waste in this class requires disposal at greater depths - tens of metres to a few hundred metres

High level waste
Waste with levels of activity concentration high enough to generate significant quantities of heat by the decay process
Waste with large amounts of long lived radionuclides
Disposal in deep, stable geological formations

Figure 3. Classification of solid radioactive waste based on radiation level.

In Table 2 are listed clearance levels for some radionuclides used in Nuclear Medicine for solid wastes release in some countries and European Union. As one may notice, almost are based on IAEA clearance levels. In the European Union the clearance levels are more restrictive and they established different scenarios to apply this definition based in their experience and practice. Most of radiological commissions around the world consider many scenarios based in the effective annual dose for public individuals less or up to 10 μSv and the collective radiological less or up to 1 man.Sv impact to estimate clearance levels (IAEA, 2009). These were established for the low probability events leading to higher radiation exposures, an additional criterion was used, namely, the effective doses due to such low probability events should not exceed 1 mSv in a year (IAEA, 2004).

Solid Radioactive Wastes in Nuclear Medicine 83

Table 2. Comparison between the clearance levels in different places and Commissions on Radiation Protection around the world

	Clearance Levels for Solid Waste (Bq/g)				
Radio-isotopes	European Union Commission[a]	German Commission[c]	Brazilian Commission[d]	Australian Commission[e]	IAEA[f]
^{18}F	1	10	10	10	10
^{131}I	1	100	100	100	100
^{177}Lu	10	1000	1000	1000	100
^{201}Tl	10	100	100	100	100
^{133}Xe	0.1150[b]	1000	1000	1000	100
^{99m}Tc	100	100	100	100	100
^{67}Ga	NF	100	100	100	100

[a] https://ec.europa.eu/energy/sites/ener/files/documents/157.pdf – rounded values derived from scenarios. NF – not found in this report

[b] Value based on activity ratio on parent nuclide, in this case ^{131}I

[c] https://www.gesetze-im-internet.de/strlschv_2001/BJNR171410001.html

[d] http://appasp.cnen.gov.br/seguranca/normas/pdf/Nrm801.pdf - values for quantities up to1000 kg

[e] https://www.legislation.gov.au/Details/F2008C00651.

[f] www-pub.iaea.org/MTCD/publications/PDF/Pub1578_web-57265295.pdf

Radiation monitoring should be conducted routinely to determine the external radiation levels and surface contamination levels inside the waste storage facility, along the boundaries of the storage facility and on the surface of waste packages (IAEA, 2006).

In surveying a waste package, independent measurements are usually made to determine dose rate (mSv/h), or activity (Bq) measured at a specified distance from the container; the radioactivity concentration of the waste content (Bq/g) or any radioactive contamination on the surfaces of the package.

The measurement of the radioactivity from the waste is important for handling the package and verifying records. The initial activity measurement provides information on the level of radioactivity present in order to determine further treatment requirements. A second measurement is required where waste is to be released at clearance levels. Gamma radiation with energies above 100 keV is easily detected allowing rapid and reliable measurements. The equipment is well suited for monitoring of low level, solid wastes from laboratories and medical use of radionuclides. Portable commercial instruments with sensitive detectors adapted to the radiation characteristics of the waste are useful but do not always allow the estimation of residual activity in numerous waste configurations (IAEA, 2000).

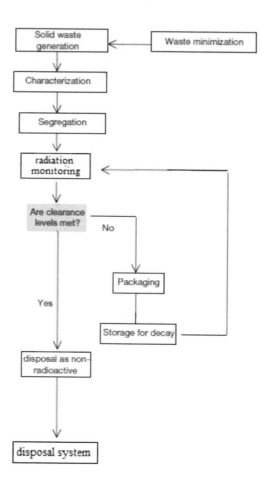

Figure 4. Flow diagram illustrating the steps in solid radioactive waste management.

A decay storage period of ten half-lives will reduce the initial radioactivity to less than one thousandth of its original radioactivity (in this case the final activity is 1/1024 of the initial activity), which in many cases means below the clearance levels for release, depending on the local regulatory requirements (IAEA, 2000). Decay storage to clearance level is almost always the preferred waste management option, both scientifically and economically. Certain categories of biohazard radioactive waste which have been subjected to pretreatment, so that they are no longer a biological hazard, can be disposed of at clearance levels. All relevant regulations for disposal of waste below the clearance levels with municipal refuse must be followed. Although, the storage period of ten half-lives it is used for many nuclear medicine facilities, a study showed that as a general rule, the ratio of residual activity to initial activity in flasks used could vary between 2% and 10% (Alabarse, et al., 2009). This percentage is currently affected by the technique adopted for the radiopharmaceutical extraction and manipulation efficiency. This study suggests that, in the absence of measuring equipment with the capacity to detect very low radiation levels or when reliable statistical data is not available, the adoption of the hypothesis that 10% of the initial activity of the liquid remains in the solid waste provides an expedite way to estimate the storage time for decay, allowing the subsequent release of solid waste in the urban conventional land fill disposal system (IAEA, 1996). It has also been pointed out that the generally adopted decay period of ten half-lives for the storage of radioactive waste before clearance only applies when the initial concentration is in order up to 1000 times the established concentration limits (Alabarse, et al., 2009).

In the Figure 4 is presented a summary flow diagram about the solid waste management in a Nuclear Medicine facility.

As part of the solid waste management program it is necessary to collect data from each package and specific facility to allow the follow up of waste from origin to their disposal. Below are listed the most important records that need to be find in a solid waste package/box and in the solid waste form records. It may vary between States according the regulatory control establishment.

- All physical and biological (if applicable) characteristics of the waste generated and radionuclide or mixture of the radionuclides present.
- Place of generation and storage (institution – nuclear medicine facility and place of storage).

- Number of package or box that contains the waste to keep follow-up.
- Responsible for the radioactive waste management program and radiation protection.
- Estimated or measured activity and dose rate of the waste package and distance from each measurement made (e.g., surface, 100 cm from surface).
- Date of dose rate and activity measured in the storage.
- Weight of waste package.
- Date of expected release.
- Certification of clearance levels and radiation monitoring.
- Signalization of radiation and biological hazard (when applicable) standard stamps.

As commented before, portable commercial instruments with sensitive detectors adapted are useful but do not always allow the estimation of residual activity in the waste package. The Brazilian commission advises in the normative (CNEN, 2014) to calculate storage time based on 10% of initial activity presented in the materials included in the waste package or an experimental trustworthy method. Usually it is defined in the beginning of each facility practice the storage time calculations based on 10% of total activity per radionuclide per week (worst case) and clearance levels to disposal based on member state regulations, in this case the Comissão Nacional de Energia Nuclear/National Commission of Nuclear Energy – CNEN in Brazil. Furthermore, people from staff usually use the Geiger Müller (GM) detector (almost of the time is the unique available detector in the facility for this kind of measurements) to measure dose rate or exposure rate, before and in the release/disposal process. Based on these records they estimate the residual activity in the waste applying the theoretical point-source relationship with the exposure/dose rate measurement.

It is already known that GM detectors are not recommended for activity measurements or its estimations (Iwahara, et al., 2002; Iwahara, et al., 2009 and Correia, et al., 2012) and this critical review is going to appoint the characteristics that are going to reinforce this affirmative. Moreover, it is

Solid Radioactive Wastes in Nuclear Medicine

being presented some data to show that this measurement and the point-source considerations are not the best scenario to use in Nuclear Medicine practice to evaluate activity for solid waste management.

PARAMETERS THAT MAY INFLUENCE THE ESTIMATION OF ACTIVITY

The GM is largely used in Nuclear Medicine to the daily routine (Ranger, 1999) and Prekeges (Prekeges, 2017) in his continue education paper reports about pitfalls in the use of radiation detection instruments. In his paper (Prekeges, 2017) is pointed out, among other problems, the relationship between the geometrical consideration of the measurement and the applicability (or not) of the inverse square-law, being commented its limitations. The GM may be used on management processes of solid radioactive waste to the radiation level (Ferreira, 2009), dose or exposure rate measurements, and also in some facilities in Brazil this measurements are used for estimation of the waste package residual activity (Ferreira, 2009). The activity estimation of the radioactive waste, commonly performed at these facilities, is based on theoretical deterministic formulae and on the relationship between exposure rate (measured from GM) and the activity. In a general way one may calculate the $\dot{X}$, based on the activity (A) and distance between the point-source and the detector (d) to a specific radioisotope characterized by its exposure rate constant (Γ_δ) by using the relation presented in Equation 1.

$$\dot{X} = \frac{A \times \Gamma\delta}{d^2} \qquad \qquad \text{Equation 1}$$

This estimation is based on a simple model considering a point-source exposure rate of R/h for 1 Ci of activity at 1 meter of distance that could be expressed as exposure rate constant - Γ_δ - in R.cm^2/h.mCi. The main objective of this topic is to explore the limitations using this relationship

(Equation 1) considering, as example, the geometry for solid radioactive waste residual.

This equation is used to estimate the activity of a solid radioactive waste by using exposure rate measured in specific recommended geometrical configuration. Some countries are allowed to dispose the radioactive waste of medical facilities based on calculation or in estimations based in exposure rate measurements (CNEN, 2014). In Brazil, is recommended to perform the dose rate measurements at surface and 100 cm far from the waste package surface and the detector's surface in the solid waste management for storage procedure (CNEN, 2014). It is easy to observe that in these geometrical configurations the point-source approximation is not appropriated for activity estimations.

In Germany for example, it is not allow disposal just based on estimative calculations. Furthermore, it is mandatory measure samples from the radioactive waste (e.g liquid state) or the entire package, in case of solid waste, measure the activity per gram and check if the clearance level is already achieved (Federal Ministry of Germany, 2017). In Figure 5 there is a scheme about solid waste management exemplification used in Germany.

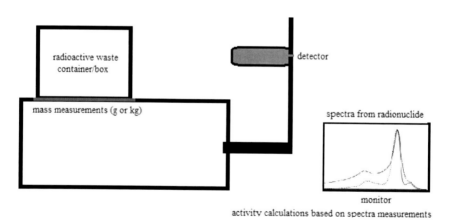

Figure 5. Example about the solid waste management in Germany to measure activity based on radionuclide gamma spectrum.

It is known that this relation, presented in Equation 1, is extremely limited because it is based on the point-source with isotropic emission approximation as a primary assumption. It does not consider the radiation absorption and scatter and it neglects the distribution of the radionuclide in the package of waste. Therefore the GM detectors are not recommended to estimate activity and it has not being recommended to use for patients measurements in case of activity administrations (Iwahara, et al., 2002; Iwahara, et al., 2009 and Correia, et al., 2012). The main idea of this topic is to estimate the impact of using the Equation 1 on daily routine to estimate the activity of solid radioactive waste and consequently the impact on storage time. To explore some of these limitations we used Monte Carlo technique, deterministic calculations and few experimental data. This topic will present the situations listed below:

(i)   Compare the experimental exposure rate by using GM to the exposure rate calculated based on activity measured by a dose calibrator;

(ii) Use Monte Carlo simulations to explore the difference on considering a non-point-source as a point-source;

(iii) Use the geometrical corrections to explore the impact of the point-source consideration to a real geometry of solid waste storage on activity estimation and on storage time of the radioactive waste.

THE COMPARISON BETWEEN EXPOSURE RATE MEASURED BY USING GEIGER MÜLLER AND THE EXPOSURE RATE CALCULATED FROM ACTIVITY BY CONSIDERING THE SOURCE AS A POINT-SOURCE

The experimental setup, presented in Figure 6, was composed by a GM detector – GM - from Fluke - Victoreen manufacture, ASM990S model with area probe model 491-40, well calibrated to the conditions of the measurement (Elimpex, 2013) and standard sources and other radioisotopes commonly used in Nuclear Medicine field. The GM was used to measure the exposure rate. The standard sources selected to the measurements were the ones usually applied on dose calibrator quality control in Nuclear Medicine from Eckert & Ziegler Isotope Products (Eckert and Ziegler, 2013) manufacture: Barium 133 (Ba-133), Cobalt 57 (Co-57) and Cesium 137 (Cs-137). All standard sources are cylindrical sources (of radius of 3.1 cm and 2.9 cm height) incorporated in an epoxy matrix (of density 1 g/cm^3) with a plastic covering. To compare the radioisotopes commonly used on Nuclear Medicine field is was used as examples Gallium 67 (Ga-67, cylindrical source of radius 0.9 cm and 1.1 cm height), Technetium 99m (Tc-99m, cylindrical source of radius 0.4 cm and 0.7 cm height) and Iodine 131 (I-131, cylindrical source of radius 0.4 cm and 0.6 cm height). The experimental data were collected in a Nuclear Medicine facility, in a low background (BG) room which was subtracted from all measurements.

The experimental setup had a GM detector face centered on the standard source base with detector–source distance of 30 cm being conducted thirty measurements for each standard source considering 30 seconds (t$_{30}$) of accumulated exposure (**X**). The exposure rate ($\dot{X}$) was calculated dividing **X** by t$_{30}$.

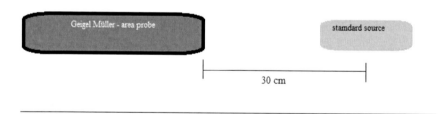

Figure 6. Scheme of the geometry used to collect the experimental data with GM and standard source to validate exposure rate measurement in the simulation.

The methodology to calculate the exposure rate to each condition is described by Brambilla and Hoff (Brambilla and Hoff, 2014). The GM and the dose calibrator were calibrated and all corrections coefficients recommended by the manufacturer were applied. The correction in time was performed to the measurements of exposure rate with GM and to define the activity by applying the exponential decay model (Equation 2), where the activity (A) at an instant (t_1) is dependent on the initial activity (A_o) measured at the instant t_o, the time (t) that has passed between the instant t_o and t_1 and the physical half-life of the radioisotope of interest. Table 3 presents the Γ_δ and the physical half-life used to calculate the exposure rate, presented in Figure 7.

$$A = A_0\, e^{\frac{-\ln(2).t}{t_{1/2}}} \qquad \text{Equation 2}$$

Table 3. Exposure rate constant and physical half-life used to calculate the exposure rate presented in Figure 5

Radioisotope	$\Gamma\delta\left(\frac{R.cm^2}{mCi.h}\right)$[a]	Physical half-life[b]
Cs-137	3.43	30.080 year
Co-57	0.563	271.74 days
Ba-133	3.04	10.551 year
Tc-99m	0.795	6.02 hours
I-131	2.2	192.6048 hours
Ga-67	0.803	78.2808 hours

[a] Data from Smith and Stabin (2012).
[b] Data from NuDat 2.6 (at http://www.nndc.bnl.gov/nudat2/chartNuc.jsp).

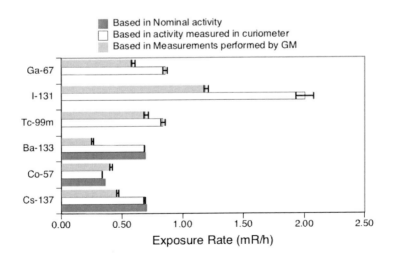

Figure 7. Exposure rate (mR/h) to six different sources comparing the measurements performed by GM (light grey column), the exposure rate estimated considering the point-source approximation to nominal activity (dark grey column) and activity measured with calibrated curiometer/dose calibrator (white column).

Figure 7 shows the average exposure rate measured to 30 measurements by using GM, the exposure rate calculated by using the point-source approximation based on activity measures by dose calibrator and, to standard sources, the exposure rate calculated by using the source point approximation based on nominal activity of the source. The error bars associated to the mean values of exposure rate are the uncertainties inferred from repeated measured values (uncertainties type A), the ones associated to the simulation are due the statistical fluctuation on application of Monte Carlo technique.

It is easy to observe that calculations made by using Equation 1 estimated the exposure rate based on nominal and measured activity always presented the closest values. The percentual differences observed between exposure rates calculated based on nominal activity and measured activity were: 2.3% to Cs-137, 8.8% to Co-57 and 1.5% to Ba-133. The dose calibrator was calibrated and the differences presented between the exposure rates majorly reflects the accuracy observed on dose calibrator quality control tests, since the calculations and parameters to estimate the exposure rate applied to the activities were the same to the same radioisotope.

The percentual differences calculated between exposure rate measured by using GM and calculated based on measured activity by the dose calibrator were: -32.4% to Cs-137, 21.6% to Co-57, -62.4% to Ba-133, -16.5% to Tc-99m, -40.4% to I-131 and -30.7% to Ga-67. It is evident the disagreement between the exposure rate presented by activity measurements made with dose calibrator and made by using GM. These differences cannot be simply justified due the conversion efficiency or geometrical limitations of the GM sensitive volume solid angle, since there are a lot of conversions factors used on these calculations that depends on different assumptions. This methodology, dependent on the GM measurements and the use of Equation 1, is used in daily routine by many facilities, and one should be aware of its limitations when using it. One commonly used parameter is the Γ_δ. The Equation 3 presents the estimation of this quantity defined to an isotropic point-source, being dependent on the spectra measurements (of photon emission energy E_i and yield f_i to each radionuclide) and mass-

Solid Radioactive Wastes in Nuclear Medicine

energy absorption coefficient $\left(\frac{\mu_{ab}}{\rho}\right)_i$ in air for photons E_i (Lauridsen, 1982 & Smith and Stabin, 2012). In general the combined emissions for parent/progeny are not taken into account.

$$\Gamma_\delta = \frac{1}{4\pi} \Sigma_1^n E_i f_i \left(\frac{\mu_{ab}}{\rho}\right)_i \qquad \text{Equation 3}$$

Table 4 shows values of Γ_δ calculated by different references to the radioisotopes used in experimental setup presented in this review.

Table 4. Exposure rate constant in $\left(\frac{Rcm^2}{mCi.h}\right)$,

presented by different authors

Radioisotope	NCRP# 491976[a]	Lauridsen 1982[b]	Attix 2004[c]	Atiix 2004[d]	Bushberg 2002[e]	Bushberg 2002[f]	Hebrion Filho 2004[g]	Smith 2012[h]	Khalil 2017[i]
Cs-137	3.2	3.224	3.2	3.249	3.25	3.25	3.3	3.43	NA
Co-57	NA	0.583	NA	NA	0.56	0.56	NA	0.563	NA
Ba-133	NA	2.024	NA	NA	NA	NA	NA	3.04	NA
Tc-99m	NA	0.590	NA	NA	0.62	0.60	NA	0.795	0.722
I-131	NA	2.158	NA	NA	2.18	2.15	2.2	2.2	2.128
Ga-67	NA	0.788	NA	NA	0.75	0.75	NA	0.803	1.036

[a] Table 28 in page 89.

[b] Table 4, photons of energy below 30 keV were excluded and from X-rays were omitted.

[c] Table 6.1, Γ_δ calculated by L.T. Dillman from decay-scheme data, assuming W_{air} = 33.70 eV, photons of energy below 11.3 keV were excluded.

[d] Table 6.1, Γ_δ calculated by L.T. Dillman from decay-scheme data, adjusted downward assuming W_{air} = 33.97 eV, photons of energy below 11.3 keV were excluded.

[e] Table 23-14, photons of energy below 20 keV were excluded.

[f] Table 23-14, photons of energy below 30 keV were excluded.

[g] Table I-10 in page 48.

[h] Table 1, photons of energy below 15 keV were excluded (limit of buildup factors used). Bremsstrahlung was neglected. Mass-energy absorption coefficients were obtained by log-log interpolation of Hubbell and Seltzer (1996). Exposure rate constant used in the conversion of activity to exposure rate presented in Figure 2.

[i] Table 1.4, photons of energy below 20 keV were excluded.

Table 5. Mean energy expended by an electron to produce an ion pair in air published by different authors

W_{air} (eV)	Standard Deviation (eV)	Reference
33.97	0.05	O'Brien and Bueermann, 2009
33.79	NA	NBS#85 (National Bureau of Standards Handbook, 1964)
33.73	0.15	Wighted cited by NBS#85 Table 1 A10 (National Bureau of Standards Handbook, 1964)
33.9	0.9	Weiss and Bernstein, 1955
33.6	0.3	Gross et al. 1957 cited by NBS#85 Table 1 A10 (National Bureau of Standards Handbook, 1964)
33.7	0.3	Bay et al. 1957 cited by NBS#85 Table 1 A10 (National Bureau of Standards Handbook, 1964)
33.9	0.5	Goodwin 1959 cited by NBS#85 Table 1 A10 (National Bureau of Standards Handbook, 1964)
33.8	0.4	Reid and Johns 1961 cited by NBS#85 Table 1 A10 (National Bureau of Standards Handbook, 1964)
33.84	0.34	Myers et al. 1961 cited by NBS#85 Table 1 A10 (National Bureau of Standards Handbook, 1964)
33.97	0.05	Boutillon and Perrroche-Roux, 1987
33.85	0.05	ICRU 1979 cited by Boutillon Perrroche-Roux, 1987
35.0	0.5	Valentine, 1952
33.74	0.02	Niatel, 1977 – NBS/BIPM
33.68	0.02	Niatel, 1977 – LMRI/BIPM
33.72	0.02	Niatel, 1977 – NBS/BIPM+LMRI/BIPM
33.84	NA	Myers et al., 1961

Taking the data of Smith and Stabin (Smith and Stabin, 2012) as reference one can observe that for Cs-137 the percentual differences are between -7% and -4%, which mean that the reference for present Γ_δ value larger than the presented by the other authors. To Co-57 the reference presented +4% from Lauridsen (Lauridsen, 1982) and -1% from both values presented by Bushberg (Bushberg, 2002). The Γ_δ values for I-131 presented differences between -3% and -1%. These three radioisotopes presented differences always inferior to ±10%, the limit of acceptance of reproducibility and accuracy tests performed in quality control applied to the activimeters. The Ba-133 presented value of Γ_δ with -33% of difference

Solid Radioactive Wastes in Nuclear Medicine 95

between Lauridsen (Lauridsen, 1982) and the reference. For Tc-99m the percentual differences between the reference were between -9% and -26%. Usually the percentual differences observed to Ga-67 were between -7% and -2%, with exception for Khalil (Khalil, 2017) where the difference presented was +29%. Considering the variables and its relation presented on Equation 3, the possible source for the observed similarities or differences may be due the energy spectra (probability of each energy and minimum energy considered to the spectra) and the mass-energy absorption coefficient used on calculations.

The Equation 1 may give a result in a non-usual unit $\left(\frac{MeV}{dis.m}\right)$, so one need to convert it to the usual unit $\left(\frac{R.cm^2}{mCi.h}\right)$. To convert this quantity one need to consider the mean energy expended by an electron of charge "e" to produce an ion pair (W_{pair}) in air (henceforth called W_{air}) and other conversions factor of unit. These may be additionals factors responsible for the differences or similarities presented on Γ_δ in Table 4. The unit conversion factors used has variations on its values smaller than 0.5%. Therefore, the W_{air} is another factor that could cause difference on Γ_δ estimation. The W_{air} may be defined experimentally or by empirical fitting formulae. Table 5 presents the W_{air} published by different authors from 1952 up to 2009. In the deterministic calculations and conversions of units presented in this topic was used, as reference, the value published by O'Brien and Bueermann (O'Brien and Bueermann, 2009).

Considering all data presented in Table 5 the maximum percentual difference of 4.2% was observed between Valentine (Valentine, 1952) and Gross et al. 1957 cited by NBS#85 (National Bureau of Standards Handbook, 1964). However if one exclude the data published by Valentine (Valentine, 1952) the maximum percentual difference drops down to 1.1% that shows a good agreement between most of all data published estimated with different methodologies.

Another way to estimate the W_{air} is using empirical fitting equation. Grebes (Grebes, 1935) presented an empirical formula (Equation 4) where V represents the energy of particle, in kV, and V_i is the ionizing potential of air (1.7×10^{-2} kV). This formula calculates the W_{air} considering energy lost in the whole track divided by the number of ions formed in this track and it is valid for electrons with kinetic energy from 0.3 keV up to 60.0 keV. The formula presented in Equation 5 is a differential equation of the energy lost in an element track divided by the number of ions formed in this element track.

$$W_{air} = 31.62 + \frac{5.27}{\sqrt{V-V_i}} \pm 0.08 \qquad \text{Equation 4}$$

$$W_{air}^{-} = \frac{\left(31.62 + \frac{5.27}{\sqrt{V-V_i}}\right)^2}{31.62 + \frac{7.91}{\sqrt{V-V_i}}} \qquad \text{Equation 5}$$

According to ICRU document 85 (ICRU, 1964) the mean energy expended by an electron to produce an ion pair may have the energy dependence negligible for electrons on gas to energies above 20 keV. The Figure 8 presents graphics of published data for W_{air} from different sources (particles with charge and mass of the electron and photons).

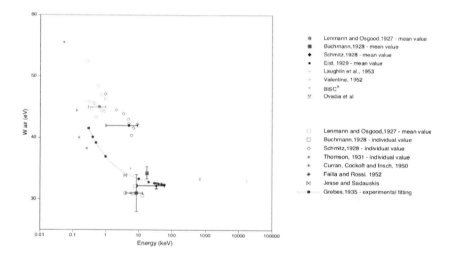

Figure 8. (Continued)

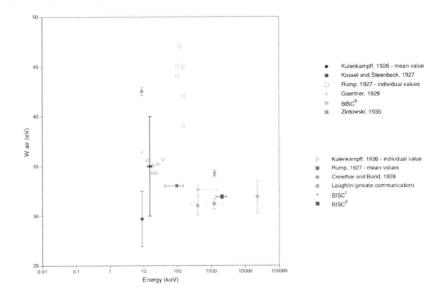

Figure 8. Mean energy expended by an electron to produce an ion pair in air from different experiments, considering as source: (above) particles with charge and mass of the electron (electrons, beta particles and cathodic rays) and (below) photons (x or gamma radiation). In this last cited configuration, the energy is expended be secondary electrons generated when photons are interacting to the matter. Filled markers represents the mean value for a group of data, empty markers represents the data for one unique energy (usually value that compose a mean value), asterisk and crosses represent individual data reported. The bars presented with few data represent the variation on group data evaluated in energy and W_{air} values.
Data source by (Binks, 1954; Bernier, 1956; Grebes, 1935).

The Figure 8 (above) presents high variation on W_{air} values for electron incident energy from 0.1 keV up to 10 keV. The general tendency is to have the W_{air} values decreasing it value with the increase on kinetic energy for electrons. The same general inverse relation between kinetic energy of the incident particle and W_{air} values may be observed on data presented in Figure 8 (below), but this tendency is not so clear. It is important to notice the energy reported on X axis is the energy of the incident photons, but the W_{air} is done by secondary electrons that present lower energy than the incident energy of the incident photons. So when one have incident photons it would be useful to estimate the mean energy of secondary electrons to associate this to the W_{air} value. Other important information is that the data presented by Rump (Rump, 1927), seems to have an average much smaller than the individual data reported in his publication. In fact Rump commented he sent the detector for calibration and discovered that his data was around 20% over estimated, however he just apply the correction to the average value presented in this paper. One may see that incident energy can influence the energy expended to from an ion pair in air.

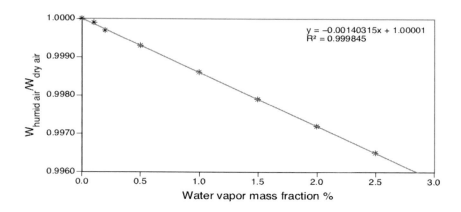

Figure 9. Correction factor for mean energy expended by an electron to produce an ion pair in air ($W_{humid\ air}/W_{dry\ air}$) in the humidity correction for the NBS-NIST spherical graphite ionization chambers (Data from Stephen and Bergstrom, 2003 – Table 12, page 374).

Besides the energy and the incident particle one may explore another factor to be evaluated for measurements performed, especially for measurements free in air or by using not pressurized sensitive volume, the humidity of air may have impact in the results. It is important to clarify that this characteristics are related to the calculation of Γ_δ, because the GM has pressurized sensitive volume. The Figure 9 shows the behavior of the W_{air} for humid air for the NBS-NIST spherical graphite ionization chambers.

It is known that in ion chambers with sensitive volume free in air the volume recombination rate increases with the increase on humidity level (Knoll, 2013), which results in a decrease in number of ion pair formed in the gas. Looking to the graphic presented in Figure 9 one may think that the air humidity is not a relevant parameter, because the variation presented is inferior to 0.004. However one need to take into account the in tropical

Solid Radioactive Wastes in Nuclear Medicine 99

countries usually the air humidity are much larger than the presented in this graphic, being usually higher than 70%. According to Andreo et al. (2005) "It is known that the (W_{air}/e) value for air at a temperature of 20°C at pressure of 101.325 kPa and 50% relative humidity is 0.6% lower than that for dry air at the same temperature and pressure, resulting in a value of 33.77 J/C instead of 33.97 J/C."

Another important parameter to take into account in the calculations of exposure are the mass-energy absorption coefficient used and the minimum energy of the spectra considered on Γ_δ calculations. It would be interesting to get the minimum energy of the spectra considered closer to the nominal radiation detection limit of the GM used. For the exposure rate calculated in this topic published Γ_δ by Smith and Stabin (Smith and Stabin, 2012) were selected because it exclude photons of energy below 15 keV that is in agreement with the specifications of the GM manufacturer of measure gamma above 12 keV (Elimpex, 2013) and the mass-energy absorption coefficients used were obtained by log-log interpolation of Hubbell and Seltzer data (Smith and Stabin, 2012) that is validated.

Besides the considerations on calculations one need to know the specific characteristics of the used GM such as the energy dependency and the geometrical corrections applied to the measurements. The sign 7collected by the GM is independent on the energy of the incoming photon due the amplification effect of the gas (Ranger, 1999) associated to the difference of potential applied to collet the generated charge. However, its probability of photon detection is dependent on incoming photon energy and on the total mass attenuation coefficient to the gas of the sensitive volume. It is important to correct the energy dependency because the GM is incapable of discriminating the photons energies that interact to its sensitive and consequently incapable of correct it. Figure 10 shows the energy dependency of the Victoreen GM area probe model 491-40 (used for the experimental data measured by GM, presented on Figure 6) and the energy dependency correction factor (EDCF) to each radioisotope used in the consideration of this topic. To select the proper energy correction factor one need to calculate the effective energy of the spectrum, by calculation the half-value layer (HVL) and use the relation $\mu = \dfrac{\ln(2)}{HVL}$ to estimate the mass attenuation coefficient. Then one may use data of total mass attenuation coefficient to aluminum to get the effective energy.

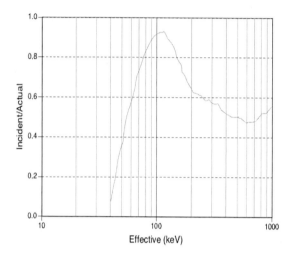

Figure 10. Victoreen GM area probe model 491-40 energy dependency, data extracted from Elimpex, 2013 to be used on data correction with probe of beta shielding closed, and the energy dependency correction factor applied to the radioisotopes use as examples in this chapter.

By observing the graphic in Figure 10 one can see that the measurement of GM (named as Actual in the Y-axis) is over estimated since the ratio of the estimated incident beam (named as Incident in the plot) by Actual is always smaller than 1. So, if one analyze the graphic presented in Figure 7 (the presented the exposure rates measured by GM with no EDCF applied), one may think it is wrong. Since if is applied the EDCF the presented exposure rate will enlarge the difference, by reducing even more the average exposure rate values. However, there are other correction factors that should be applied considering the angular efficiency, by correcting the solid angle of GM collection (generally small to a real geometry) in relation to the solid angle of emission source. This quantity is dependent of the experimental setup geometry and considering non point-sources it can be very difficult of estimate. In the geometry used in this topic, considering the point-source with isotropic gamma emission, the solid angle of GM collection was about

0.0064 sr compared to the 4π sr of emission. There is another factor that is complicate to be introduced in this analysis is the geometrical condition of experimental setup such as the increase in counts in the longitudinal volume direction of the GM sensitive volume in addition to the intrinsic characteristics of the detector (absolute efficiency and observe the additional limitations in the GM manufacturer manual). Despite of the difficulties on apply all corrections factors and the factor that it not present accurate exposure rate to a large energy range, the GM is largely used in Nuclear Medicine specially to survey application because it hight sensitivity (approximately 10 times when compared to other ionizing detector) (Ranger, 1999). Another important characteristic is that GM have relative long dead time (usually hundreds of microseconds) because it operates in pulse-mode then it cannot be used for survey in high level radiation filed (Knoll, 2013 and Ranger, 1999).

In this context, the consideration is the mean energy expended by an electron of charge "e" to produce an ion pair in gas is an important factor to be considered. The GM is a pressurized gas chamber and in this case one needs to use the mean energy expended by an electron of charge "e" to produce an ion pair in neon gas (W_{Ne}) or in helium gas (W_{He}). The Figure 11 presents W_{Ne} and W_{He} published by different authors from 1955 up to 2015.

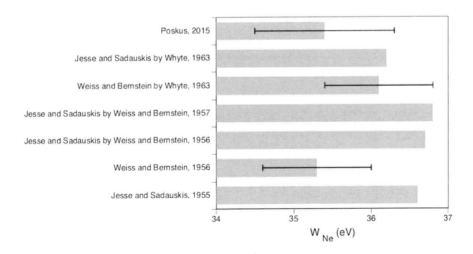

Figure 11. (Continued)

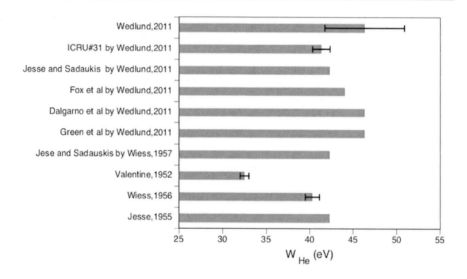

Figure 11. Mean energy expended by an electron to produce an ion pair in neon gas (above plot) and helium gas (below plot) published by different authors.

Both gases presented above have different sensitive volumes to give a general idea that it can presents variations on data measurement. However, the main gas of interest in this Chapter is the neon gas (used in the detector simulations), because helium gas is the quenching gas in the GM detector. The maximum percentual difference presented among all data for neon gas was 4.3%, between data published by Weiss and Bernstein (Weiss and Bernstein, 1956) and Jesse and Sadaukis (Weiss and Bernstein, 1957), and for helium gas was 142.5%, between Valentine (Valentine, 1952) and data published by Wedlund (Wedlund, 2011). Even for neon gas the percentual differences presented in the mean values were larger than the uncertainties presented to each value. The differences on W_{Ne} and W_{He} may be due to the difference on methodologies, purity of the hellium gas and the gas pressure. On calculation for GM detector one need to consider the W_{Ne} to estimate the number of ion pair generated in sensitive volume.

In conclusion, one may see that besides the point-source approximation equation limitations, to use it properly to estimate the activity of the solid waste one need to pay attention on different correction factor and parameters

Solid Radioactive Wastes in Nuclear Medicine

of the GM to get the proper measurement. Some of these parameters are not easy to apply in daily routine. However, the impossibility of determination of some corrections does not exempt one of have the knowledge of the technique limitations. It is clear that the above cited parameters in calculation and detection results, usually, in under estimation of the exposure rate and consequently the misestimation on solid waste activity associated to this exposure rate. It should be interesting when apply this calculations one be careful about the selection of: Γ_δ (considering different parameters in its calculation such as: minimum energy considered in the gamma spectra, W_{air} and used on conversion of units); about specification of the GM and use the GM probe energy dependency and angular limitations to correct its measurements or, at least, understand it systematic deviation on the measurements.

THE INFLUENCE OF POINT-SOURCE ASSUMPTION ON ACTIVITY ESTIMATION AND STORAGE TIME FOR SOLID WASTE GENERATED IN NUCLEAR MEDICINE

In this topic it will be explored, by using Monte Carlo technique, the influence on considering the solid radiative waste as a point-source. To perform the Monte Carlo simulation it was used Geant4 version 9.1.p03 (Agostinelli et al., 2003 and Allison et al., 2006). The Geant4 version 9.1.p03 application development and additional information may be found in (Brambilla and Hoff, 2014).

The GM simulated was described as a detector sensitive volume with the characteristics of Fluke - Victoreen ASM990S connected to an area probe 491-40 model (Elimpex, 2013), consisting of a cylinder with external dimensions of 1.35 cm radius and 2.01 cm of length. The housing was defined as stainless steel with thickness of 0.05 cm (Karaiskos et al., 1998). The internal active volume of 10.141 cm^3 also simulated cylindrical, with a radius of 1.30 cm and a height of 1.91 cm consists of pure neon (Ne) with a density of 8.385.10^{-1} g/m^3 under a pressure of 1.021 atm.

Two distinct geometries were simulated to exemplify the influence of geometry of experimental setup on exposure rate: 1) point-source and 2) box source geometry (Descarpack®, henceforth named active packed). Both simulated sources presented the same spectrum and activity. The spectra were simulated according definition of gamma emissions for each radioisotope from NuDat2.6 (National Nuclear Data, 2017), neglecting the energies bellow 12 keV as recommended by the limitation on radiation detection defined (Elimpex, 2013). These two geometries were developed to assess the impact on exposure generated by waste active packed, applied on radioactive waste management, which are used to geometries defined in clinical practice on many Nuclear Medicine facilities in Brazil. Each geometry has been explored in its particularities.

The MC method was based on Geant4 version 9.1.p03 implementation by evoking standard library to model the radiation transport in the simulated medium, considering limit to produce secondary particles of 0.01 mm (electrons and photons). Additional information about the simulation and its validation of Monte Carlo physics models are presented by Brambilla and Hoff (Brambilla and Hoff, 2014). The GM detector quenching gas was not considered in the simulation, since it has the function of controlling the ionization cascade detection and enabling new decreasing the detector's dead time (Knoll, 2013) and this factor does not directly affect the simulation results of absorbed dose in the detector generated from the neon gas ionization. The simulations results were normalized by the number of photons emitted from the source, so as to represent the probability of photon energy absorption, which is only valid for the studied geometry.

The point-source implemented, in the Geant4, was set with an isotropic emission placed at center of the active packed. The active packed (Descarpack, 2012) was simulated with dimensions (30.5 x 31.5 x 25.0) cm^3. The non-punctual source considered the photons emission in the active packed homogeneously distributed in its volume with probability of angular emission isotropic. Both (point and non-punctual) geometries were simulated considering different positions of the GM always centered on the active packed surfaces:

1. distance between the detector's surface and source center at 10 cm,
2. distance between the detector's surface and source center of 100 cm,
3. distance between the detector's surface of 100 cm and source center.

To perform the comparison the simulated exposure was normalized by emitted photon by the source (representing the average absorption energy probability by sensitive volume).

The relative impact to the geometry (RIG) was defined as Equation 6 represents the relationship between the probability of deposited energy in the detector by photons emitted by active packed ($Prob_{E_{ab}|active\ packed|}$) and the probability of the energy being deposited in the detector per photon emitted by the point-source ($Prob_{E_{ab}|point|}$). The only factor that was changed in the simulation was the source description and the relation to calculate RIG will reflect the effect of this change.

$$RIG = \frac{Prob_{E_{ab}|active\ packed|}}{Prob_{E_{ab}|point|}} \qquad \text{Equation 6}$$

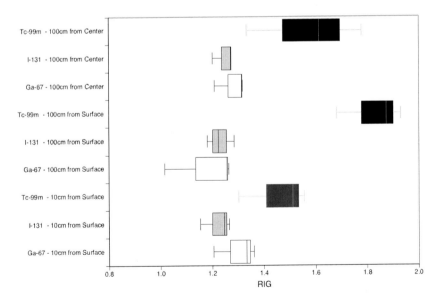

Figure 12. Box plot of the RIG calculated to different geometries and radioisotopes used in Nuclear Medicine.

The Figure 7 presents a box plot of RIG calculated to different radioisotopes considering the 3 possibilities of geometries and the six faces (surfaces) of the active packed. One need to remind that the point-source geometry considered the same geometry as the non-point-source (position of the detector), but removing the active packed and adding a point-source at the position of the geometrical center of the active packed.

By looking in Figure 11 one can see that the non-punctual geometry in fact may overestimate the activity and the exposure rates in average values between 20% and 70% depending on the geometry of data collection and radioisotope. The larger fluctuations and differences were observed for Tc-99m which can be explained because this radioisotope presents the lower effective energy (140.51 keV) compared to effective energy the presented by to other two studied radioisotopes (I-131 with 419.93 keV and Ga-67 with 265.94 keV), being in this way more susceptible to the absorption and scattering influences of the air and the casing of the active packed. Since the active packed is a parallelepiped and not a cube, each condition presented three distances from the center of it (distance used to calculate the activity in the point-source approximation methodology). Table 6 presents the RIG considering each dimension X, Y and Z of the active packed and this distance between the surface of the detector and the position of the geometrical center of the active packed.

Table 6 presents RIG values always larger than 1, which confirm the over estimation of the exposure rate, and consequently, of the activity estimation when the active packed is treated as a point-source. One may observe the variation on RIG value as function of the radioisotope and irradiation geometry. Always Z position (lines 3, 6 and 9) presented lower value of RIG because in this axis there is larger thickness of absorbers that represents the locking system of the active packed. As larger as the effective energy of the radioisotope lower is the RIG value, it is due the reduction of the probability of scattering and absorption by air and the active packed.

Solid Radioactive Wastes in Nuclear Medicine

Table 6. RIG and its absolute variation (ΔRIG) to different irradiation configuration: distance between the detector's surface and source center of 10 cm on lines 1 to 3, distance between the detector's surface and source center of 100 cm on lines 4 to 6, and distance between the detector's surface of 100 cm and source center on lines 7 to 9. The order of presentation for the average values of RIG follow axis X, Y and Z

Distance (cm)	Ga-67		I-131		Tc-99m	
	RIG	ΔRIG	RIG	ΔRIG	RIG	ΔRIG
25.25	1.332	0.005	1.246	0.003	1.513	0.003
25.75	1.360	0.005	1.266	0.003	1.556	0.003
22.50	1.205	0.005	1.155	0.003	1.303	0.003
100.00	1.315	0.006	1.273	0.003	1.778	0.067
100.00	1.318	0.006	1.275	0.003	1.613	0.053
100.00	1.210	0.006	1.201	0.003	1.334	0.041
115.25	1.256	0.011	1.225	0.009	1.873	0.273
115.75	1.264	0.012	1.286	0.012	1.930	0.262
112.50	1.016	0.008	1.180	0.009	1.682	0.251

In a general consideration about all the influences on may observe comparing the estimation of activity based on exposure rate measured by GM (accuracy) and the influence on geometry, to the evaluated cases in this chapter on may conclude that: (i) -30.7% in accuracy and average RIG to each geometry about +30% for $RIG_{10cm\_surface}$, +28% for RIG_{100cm_center} and +16% for $RIG_{100cm\_surface}$ to Ga-67, -40.4% in accuracy and average RIG to each geometry about +22% for $RIG_{10cm\_surface}$, +25% for RIG_{100cm_center} and +22% for $RIG_{100cm\_surface}$ to I-131, -16.5% in accuracy and average RIG to each geometry about +45% for $RIG_{10cm\_surface}$, +57% for RIG_{100cm_center} and +84% for $RIG_{100cm\_surface}$ to Tc-99m. The radioisotope with higher effective energy (I-131) presented lower fluctuation on RIG value. The radioisotope with lower effective energy (Tc-99m) presented the highest fluctuation on RIG value, with a tendency of increase its value with the increase on distance between detector and active packed. And the Ga-67 which presented a significant difference on RIG value and a tendency of reduce this value with increase on distance. The opposite tendency on RIG values with distance presented by Tc-99m and Ga-67 may be justified by its effective energy 140.51 keV and 265.94 keV, respectively. The effective energy for Tc-99m presents more susceptibility and larger mass attenuation coefficient for air and active packed material (when compared to Ga-67 or I-131) which means the Tc-99m photons have larger probability of interact being scattered and absorbed.

The results presented here makes no possible to define one unique strategy for the calculations and corrections to be applied, but it can prove that each case/geometry of irradiation and GM model needs to be evaluated carefully.

Moreover, it is presented a study case to show activity estimation and the impact in storage time based on simulations data and considering 10% of initial activity authorized in a Nuclear Medicine facility. This hypothetic facility is allowed acquire the amount of activity per week listed below:

- Iodine 131: 2.78 GBq
- Technetium 99m: 37 GBq
- Gallium 67: 0.37 GBq

In each case were used the consideration based on Equation 1 to calculate the source activity based on rate exposure from simulations data and then this data were compared with 10% of initial activity for storage time estimative using the case of sources listed above.

The storage time based on clearance levels (listed in Table 2) was calculated based on Equation 7 for each clinical source and for each case already presented in this Chapter (100 cm distance between the GM and the center of the source, 100 cm distance between the GM and the source surface and 10 cm between the GM and the source surface).

$$t_{storage} = \left[-\ln\left(\frac{A_c}{A_0}\right)\right] \times \left[\frac{t_{1/2}}{-\ln(2)}\right] \qquad \text{Equation 7}$$

Where: $t_{storage}$ is the time to allow the waste package release, A_c is the clearance level in Bq/g established for each source, A_0 is the initial activity

Solid Radioactive Wastes in Nuclear Medicine 109

calculated theoretically using Equation 1 and considering for packed the mass of 29.06 g (according to the geometry in the simulation data based on Descarpack), exposure rate cases already simulated and described above for each source and specific case, and $t_{1/2}$ is the physical half-life from each source according to Table 3.

Furthermore, it will be presented the effect of apply the geometry corrections for each source package and it impacts in time storage.

Table 7 presents for ^{131}I, ^{99m}Tc and ^{67}Ga the activity calculations based on Equation 1 and rate exposure from MC simulations in each case. The storage time can be notice and compared with theoretical conservative consideration of 10% initial activity per week.

Based in the results of activity estimative and time release presented is possible to notice that activity for 100 cm detector center and surface from these sources are underestimated (based on known/true activity), on the other hand for 10 cm of distance to the detector the activities were overestimated compared to the point-source estimative. Like it was already discussed in this chapter, the GM detector is not recommended to estimate activity from a source, there are many factors that affected this measure and estimative. In addition, to treat the packed as a point-source overestimate the activity compared to the point-source measurement. It is important to evaluate the release time impact for sources with short half-lives, in this example Technetium-99m that suffers low impact for all cases, however for Iodine-131 which is the higher half-life in this example presents major variations. In all cases presented, the conservative method of considering 10% of initial activity in the week guarantee that the release will be done with safety regarding the clearance levels for all sources and cases. Moreover, it is important to observe that optimization of geometry factor works in a correct manner establishing a good relationship with the time of point-source considerations. The best results for the geometry corrections were presented for the case that the distance between detector and source from the center was 100 cm. This can be explained according to the Equation 1 and the definition of exposure rate constant for the point-source theory.

Table 7. Comparison of time to release (clearance reached) for each clinical source based on different geometric cases and distances to the detector (point-source and packed source) and also compared to theoretical assumption of 10% of initial activity per week

Considering distance 100 cm from the detector to the source-center   Time to release (days)

Source	Point-source Activity (Bq)	Packed Source Activity (Bq)	Activity Geometry Corrected (Bq)	Point-source	Packed source	10% Activity/week	Packed source Geometry Corrected
I-131	1.25E+09	1.59E+09	1.27E+09	150	153	172	150
Tc-99m	1.01E+09	1.80E+09	1.15E+09	5	5	6	5
Ga-67	1.39E+09	1.82E+08	1.42E+08	51	52	60	51

Considering distance 100 cm from the detector to the source-surface   Time to release (days)

Source	Point-source Activity (Bq)	Packed Source Activity (Bq)	Activity Geometry Corrected (Bq)	Point-source	Packed source	10% Activity/week	Packed source Geometry Corrected
I-131	1.87E+07	4.59E+07	3.96E+07	102	112	172	110
Tc-99m	1.62E+08	3.40E+08	2.76E+08	4	4	6	4
Ga-67	3.80E+07	4.77E+07	2.91E+07	45	46	60	43

Considering distance 10 cm from the detector to the source-surface   Time to release (days)

Source	Point-source Activity (Bq)	Packed Source Activity (Bq)	Activity Geometry Corrected (Bq)	Point-source	Packed source	10% Activity/week	Packed source Geometry Corrected
I-131	4.18E+09	1.12E+10	8.68E+09	164	176	172	173
Tc-99m	5.00E+10	7.56E+10	6.20E+10	6	6	6	6
Ga-67	8.65E+08	1.15E+09	7.99E+08	59	61	60	59

It is possible to conclude from this study case:

1. For radionuclides with shortest physical half-lives the geometry, detection or considerations based on initial activity/week do not affected significantly the time of storage. In this case, the surface contamination must be indicated to evaluate release from control in addition of storage time considerations.
2. It is important to notice the variation in storage time according to the measurement distance. It indicates that a definition of the measure geometry needs to be taking into account for the facility routine. It is important to establish a constant method of measurement to the exposure rate and if is intended to estimate waste package activity this factor can influences a lot in the final result, at least 10 days according to Ga-67 and I-131 sources in some cases of these analysis.
3. Considering 10% of total activity per week to calculate the storage time is a safety margin to guarantee clearance levels, but it could extend the time of release and accumulate more radioactive waste in the facility storage.
4. More than one factor need to take into account for radioactive waste management, because it was notice that many factors could affect the activity and storage time estimative for waste package to reach clearance levels with confidence. For each case is important to evaluate methods of waste management and disposal considering a safety procedure to not impact environment and biological systems with the practice of nuclear instrumentation.

According to this Chapter it is important to compare the possibility of developed countries that works with robust and expensive detectors which are more trustful in the radioactive waste management. On the other hand, countries in development have higher difficulties (specially based on the financial investments) to acquire these detectors and need to adapt their routines according to the possibilities that they live in practice of Nuclear Medicine. Furthermore, it is possible to establish a safety routine according to this critical review and apply each case a confident methodology to not affect environment and people because the practice with radionuclides. It is known that ALARA (As Low as Reasonably Achievable) principles must be considering in all exposure practices and the optimization of exposure is mandatory for all practices and sources regarding to their.

REFERENCES

Alabarse F., Xavier, A., Magalhães M., Guerrero J., 2009. Solid radioactive waste: evaluation of residual Activity in nuclear medicine services. International Nuclear Atlantic Conference - INAC 2009 Rio de Janeiro, RJ, Brazil, ABEN ISBN: 978-85-99141-03-8.

Andreo, P., Seuntjens J.P., Podgorsak, E.B. Calibration of photon and electron beams. *Radiation Oncology Physics.* Vienna: IAEA, 2005.

Attix, F. H. *Introduction to Radiological Physics and Radiation Dosimetry.* Edited by WILEY-VCH Verlag GmbH & Co. KGaA, Weinheim, 1986; pp 1-607.

Bernier, J.P., Skarsgard L.D., Cormack, D.V., Johns H.E., 1956. A Calorimetric Determination of the Energy Required to Produce an Ion Pair in Air for Cobalt- 60 Gamma-Rays. *Radiation Research.* 5, 613-633.

Binks, W., 1954. Energy per ion pair. *Acta Radiologica* 117 suppl 85-104.

Boutillon, M., Perroche-Roux A. M., 1987. Re-evaluation of the W value for electrons in dry air. *Physics in Medicine and Biology.* 32, 2-213.

Brambilla, C.R., Hoff, G., 2014. Impact evaluation of the geometry on measurements of solid radioactive waste exposure rates in nuclear medicine. *Braz. J. Biom. Eng.,* 30, 330-340.

Bushberg, J.T., Seibert, J.A., Leidholdt, E.M., *The essential physics of medical imaging.* Lippincott Williams & Wilkins, Second Edition. 2002; pp 1-919.

Comissão Nacional de Energia Nuclear, 2014. Gerência de rejeitos radioativos de baixo e médio níveis de radiação NN 8.01. Resolução CNEN 167/14. [National Comission of Nuclear Energy. *Management of Radioactive Waste of Low and Medium Radiation Levels.* [2017/03/05]. Available from: http://appasp.cnen.gov.br/seguranca/normas/pdf/Nrm801.pdf.

Correia, A. R., Iwahara, A., Tauhata, L., Rezende, E.A., Chaves, T. O., de Oliveira, A.E., de Oliveira, E.M., 2012. Volume corrections factors in the measurement of 99mTc and 123I activities in radionuclide calibrators. *Radiol Bras.* 45, 93–97.

Descarpack. São Paulo. [2012/11/15]. Available from: http://www.descarpack.com.br/perfurocortante.htm.

Eckert & Ziegler. Eckert & Ziegler Isotope Products. Medical imaging sources: product information [2013/11/15]. Available from: www.isotopeproducts.com.

Eisl, V.A., 1929. Über die Ionisierung von Luft durch Kathodenetrahlen von 10-60 kV [On the ionization of air by cathode rays of 10-60 kV]. *Annalen der Physik.* 395.3, 277-313.

Elimpex. Survey Meters 491-40 Utility 1R GM Probe [2013/11/15]. Available from: www. elimpex.com/products/survey_meters/491-40/491-40.html.

European Commission, Practical Use of the Concepts on Clearance and Exemption – Part 1: Guidance on General Clearance Levels for Practices. Radiation protection 122. EURATOM: European Union, 2000, 40p.

Federal Ministry of Germany. Verordnung über den Schutz vor Schäden durch ionisierende Strahlen (Strahlenschutzverordnung - StrlSchV). [Regulation on protection against damage caused by ionizing radiation]. [2017/03/01]. Available from: www.juris.de.

Ferreira, F.C.L., Magalhães, C.M.S., Costa, N.V., Lima L.L., Cardoso, L.X., Cunha, C.J., Souza, D.N., Souza, S.O., 2009. Avaliação de Rejeitos Radioativos em Serviços de Medicina Nuclear no Estado de Sergipe [Evaluation of Radioactive Rejects in Nuclear Medicine Services in the State of Sergipe]. Scientia Plena. 5, 1-5.

Food and Agriculture Organization of the United Nations, International Atomic Energy Agency, International Labor Organization, OECD Nuclear Energy Agency, Pan American Health Organization, World Health Organization, International Basic Safety Standards for Protection against Ionizing Radiation and for the Safety of Radiation Sources. Safety Series No. 115. IAEA, Vienna, 1996.

Gases. Phys Rev., 98 (6), 1828-1831.

Gerbes, V.W., 1935. Über die Ionisierungswirkung von Kathodenstrahlen in Luft. Annalen der Physik. 415.7, 648-656. [On the Ionization Effect of Cathode Radiation in Air. Annalen der Physik. 415.7, 648-656].

Heilbron, F., Ferneno P.L., Xavier, A.M., Pontedeiro, E.M., Ferreira, R.S., Segurança Nuclear e Proteção do Meio Ambiente. Edited by FAPERJ &-papers: Rio de Janeiro. 2004; pp 1-316. [Nuclear Safety and Environment Protection. Edited by FAPERJ &-papers: Rio de Janeiro. 2004; pp 1-316].

International Atomic Energy Agency, Application of the Concepts of Exclusion, Exemption and Clearance. General Safety Guide No. RS-G-1.7. IAEA: Vienna, 2004.

International Atomic Energy Agency, Application of the Concepts of Exclusion, Exemption and Clearance. Safety Guide. Safety Series No. RS-G-1.7. IAEA: Vienna, 2004, 29 p.

International Atomic Energy Agency, Basic Factors for the Treatment and Disposal of Radioactive Wastes. Safety Series No. 24. IAEA: Vienna, 1967, 41 p.

International Atomic Energy Agency, Classification of Radioactive Waste. General Safety Guide No. GSG-1. IAEA: Vienna, 2009.

International Atomic Energy Agency, Classification of Radioactive Waste. Safety Series no. 111-G-1. IAEA: Vienna, 1994, 39 p.

International Atomic Energy Agency, Classification of Radioactive Waste. General Safety Guide. Safety Series No. GSG-1. IAEA: Vienna, 2009, 48 p.

International Atomic Energy Agency, Clearance levels for radionuclides in solid materials application of exemption principles Interim report for comment. IAEA TECDOC 855. IAEA: Vienna, 1996, 70 p.

International Atomic Energy Agency, Clearance of materials resulting from the use of radionuclides in medicine, industry and research. IAEA TECDOC 1000. IAEA: Vienna, 1998, 49 p.

International Atomic Energy Agency, Establishing a National System for Radioactive Waste Management. Safety Series no. 111-S-1. IAEA: Vienna, 1995, 28 p.

International Atomic Energy Agency, Fundamental Safety Principles. Safety Fundamentals. Safety Series No. SF-1. IAEA: Vienna, 2006, 21 p.

International Atomic Energy Agency, Governmental, Legal and Regulatory Framework for Safety. General Safety Requirements Part 1. Safety Series no. GSR Part 1 (revision 1). IAEA: Vienna, 2016, 42 p.

International Atomic Energy Agency, International Basic Safety Standards for Protection against Ionizing Radiation and for the Safety of Radiation Sources. Safety Series Standard. Safety Series No. 115. IAEA: Vienna, 1996, 30 p.

International Atomic Energy Agency, Management of Radioactive Wastes Produced by Users of Radioactive Materials Procedures and Data. Safety Series No. 70. IAEA: Vienna, 1985, 37 p.

International Atomic Energy Agency, Management of Waste from use of Radioactive Material in Medicine, Industry, Agriculture Research and Education. Safety Series No. WS- G-2.7. IAEA: Vienna, 2005.

International Atomic Energy Agency, Monitoring for Compliance with Exemption and Clearance Levels. Safety Report. Safety Series No. 67. IAEA: Vienna, 2012, 186 p.

International Atomic Energy Agency, Monitoring for Principles for the Exemption of Radiation Sources and Practices from Regulatory Control. Safety Guide. Safety Series No. 89. IAEA: Vienna, 1988, 23 p.

International Atomic Energy Agency, Predisposal Management of Radioactive Waste - General Safety Requirements Part 5. Safety Series no. GSR Part 5. IAEA: Vienna, 2009, 38 p.

International Atomic Energy Agency, Principles for Establishing Limits for the Release of Radioactive Materials into the Environment. Safety Series No. 45. IAEA: Vienna, 1978, 32 p.

International Atomic Energy Agency, Radiation Protection and Safety of Radiation Sources: International Basic Safety Standards. General Safety Requirements Part 3. Safety Series No. GSR Part 3. IAEA: Vienna, 2014, 436 p.

International Atomic Energy Agency, Regulatory Control of Radioactive Discharges to the Environment. Safety Guide. Safety Series No. WS-G-2.3. IAEA: Vienna, 2000, 43 p.

International Atomic Energy Agency, Safe Handling of Radioisotopes Radioactive waste control and disposal (1958 Edition), IAEA Safety Series No. 1. IAEA: Vienna, 1958.

International Atomic Energy Agency, Safe Handling of Radioisotope: First Edition with Revised Appendix I. Safety Series No. 1. IAEA: Vienna, 1962.

International Atomic Energy Agency, Safe Handling of Radioisotope (Edition 1973). Safety Series No. 1. IAEA: Vienna, 1973.

International Atomic Energy Agency, Safe Handling of Radioisotopes Health Physics Addendum. Safety Series No. 2. IAEA: Vienna, 1960.

International Atomic Energy Agency, Safe Handling of Radioisotopes Medical Addendum. Safety Series No. 3. IAEA: Vienna, 1960.

International Atomic Energy Agency, Storage of Radioactive Waste. General Safety Guide No. WS-G-6.1. IAEA: Vienna, 2006.

International Atomic Energy Agency, Storage of Radioactive Waste. Safety Guide. Safety Series No. WS-G-6.2. IAEA: Vienna, 2006, 55 p.

International Atomic Energy Agency, Technology of Radioactive Waste Management Avoiding Environmental Disposal. Technical Report Series no. 27. IAEA: Vienna, 1964, 147 p.

International Atomic Energy Agency, The Management of Radioactive Wastes Produced by Radioisotope Users. Safety Series No. 12. IAEA: Vienna, 1965.

Solid Radioactive Wastes in Nuclear Medicine

International Atomic Energy Agency, The Management of Radioactive Wastes Produced by Radioisotope Users Technical Addendum. Safety Series No. 19. IAEA: Vienna, 1966.

International Atomic Energy Agency, The Management System for the Processing, Handling and Storage of Radioactive Waste. Safety Guide. Safety Series No. GS-G-3.3. IAEA: Vienna, 2008, 69 p.

International Atomic Energy Agency, The Management System for the Disposal of Radioactive Waste. Safety Guide. Safety Series No. GS-G-3.4. IAEA: Vienna, 2008, 75 p.

International Atomic Energy Agency, The Principles of Radioactive Waste Management. Safety Series 111-F. IAEA: Vienna, 1995, 24 p.

International Atomic Energy Agency, The Safe Management of Sources of Radiation: principles and strategies. International Nuclear Safety Advisory Group INSAG-11. IAEA: Vienna, 1999, 27 p.

International Atomic Energy Agency, The Safety Case and Safety Assessment for the Predisposal Management of Radioactive Waste. General Safety Guide. Safety Series No. GSG-3. IAEA: Vienna, 2013, 151 p.

International Atomic Energy Agency, Management of radioactive waste from the use of radionuclides in medicine. TECDOC-1183. IAEA: Vienna, 2000.

Iwahara, A., de Oliveira, A.E., Tauhata, L., da Silva, C.J., da Silva, C.P., Braghirolli, A.M., Lopes, R.T., 2002. Performance of dose calibrators in Brazilian hospitals for activity measurements. *Applied Radiation and Isotopes.* 56, 361–367.

Iwahara, A., Tauhata, L., de Oliveira, A.E., Nícoli, I.G., Alabarse, F.G., Xavier, A.M., 2009. Proficiency test for radioactivity measurements in nuclear medicine. *J. Radioanal Nucl. Chem.* 281, 3–6.

Khalil, M.M., et al. (2017). *Basic Science of PET Imaging* [eBook]. Spring International Publishing.

Knoll, G.F. *Radiation Detection and Measurement.* [2017/03/15]. Available from: https://phyusdb.files.wordpress.com/2013/03/radiationdetectionandmeasurementbyknoll.pdf.

Kosako, T. Clearance Level Discussion on Solid Radioactive Waste. [2017/02/10]. Available from: http://www.irpa.net/irpa10/cdrom/ 01108.pdf. International Radiation Protection Association Conference – IRPA 10: Hiroshima, 2000.

Lauridsen, B., 1982. *Table of Exposure Rate Constants and Dose Equivalent Rate Constants.* (Risø-M; No. 2322).

Menon, S. An Implementor's Views on Clearance Levels for Radioactivity Contaminated Material. Waste management Symposium. WM'00 Tucson, February 27 – March 2, 2000.

Myers, I. T., et al. An adiabatic calorimeter for high precision source standardization and determination of W (air). No. HW-SA-2165. General Electric Co. Hanford Atomic Products Operation, Richland, Wash.; National Bureau of Standards, Washington, DC, 1961.

National Bureau of Standards Handbook. Physical Aspects of Irradiation; NBS#85; U.S. Government Printing Office:Washington, Washington D.C., 1964; Vol. 85, pp 1-106.

National Council on Radiation Protection and Measurements. Structural Shielding design and evaluation for medical use of x ray and gamma the rays of energy up to 10 MeV; Report NCRP#49. Edited by National Council on Radiation Protection and Measurements. Third Edition, 1998; pp 1-133.

National Nuclear Data Center, Brookhaven National Laboratory. NuDat2.6 [2017/03/15]. Available from: http://www.nndc.bnl.gov/nudat2/ chartNuc.jsp.

Nordic Council of Ministers. *Guidance on Clearance from Regulatory Control of Radioactive Materials* – TemaNord: Copenhagen, 1994, 107 p.

O'Brien, M., Bueermann, L., 2009. Comparison of the NIST and PTB Air-Kerma Standards for Low-Energy X-Rays. *J. Res. Natl. Inst. Stand. Technol.* 114, 321-331.

Prekeges, J.L., 2014. Sweating the Small Stuff: Pitfalls in the Use of Radiation Detection Instruments. *J Nucl Med Technol.* 42, 81–91.

Ranger, N.T., 1999. The AAPM/RSNA Physics Tutorial for Residents – Radiation Detectors in Nuclear Medicine. *Imaging &Therapeutic Technology.* 19, 481-502.

Seltzer, S.M., Paul M.B., 2003. Changes in the US primary standards for the air kerma from gamma-ray beams. *Journal of research of the National Institute of Standards and Technology.* 108, 5-359.

Smith, D. S., Stabin, M. G., 2012. Exposure rate constants and lead shielding values for over 1,100 radionuclides. *Health Physics.* 102, 271-291.

Valentine, J. M., 1952. Energy per Ion Pair for Electrons in Gases and Gas Mixtures. *Proc. R. Soc. Lond.* 211, 75-85.

Wedlund, C.S., Gronoff, G., Lilensten J., Ménager H., Barthélemy M., 2011. Comprehensive calculation of the energy per ion pair or W values for five major planetary upper atmospheres. *Ann. Geophys.* 29, 187–195.

Weiss J., and Bernstein W., 1955. Energy Required to Produce One Ion Pair for Several Gases. *Phys. Rev.* 98, 1828.

BIOGRAPHICAL SKETCHES

Cláudia Régio Brambilla
Institute of Neuroscience and Medicine, Medical Imaging Physics,
(INM – 4), Forschungszentrum Juelich GmbH,
Wilhelm-Johnen-Straße, 52428, Jülich, Germany.

Education: Phd Student in Health Science (Dr rer med)

Research and Professional Experience:

Research Experience
January 2017 PhD Researcher Fellowship: Project *"Multimodal Neuroimaging with 11C-ABP688 PET/MR/EEG"*. DAAD scholarship "Research Grants – Doctoral Programs in Germany". Institute of Neuroscience and Medicine, Medical Imaging Physics, (INM – 4), Forschungszentrum Juelich GmbH, Jülich, Germany.

November 2007 – December 2010 Fellowship Junior Researcher: Project *"Experimental environment based on grid technology for training staff resources in Nuclear Medicine"*. Grant Fondo Regional para la Inovación Digital de la America Latina y Caribe (FRIDA). Medical Imaging Research Center (NIMed), Faculty of Physics, Department of Applied and Theoretical Physics Pontifícia Universidade Católica do Rio Grande do Sul – PUCRS.

October 2009 – December 2009 Fellowship Junior Researcher: Project *"Capacity building in the production of radiopharmaceuticals with a cyclotron for clinical applications"*. Grant IAEA- A/2/016. Faculty of Physics, Department of Applied and Theoretical Physics Pontifícia Universidade Católica do Rio Grande do Sul – PUCRS.

June 2006 – July 2007 Research Assistant: Project *"Development of Protocol for Renal Absolute Quantification with 99mTc-DMSA"*. Grant CNPq.Medical Imaging Research Center (NIMed), Faculty of Physics, Department of Applied and Theoretical Physics Pontifícia Universidade Católica do Rio Grande do Sul – PUCRS

July 2005 – July 2006 Research Assistant: Project *"Quantification of nuclear medicine images using Monte Carlo Simulations of Human's Models"*. Grant FAPERGS. Medical Imaging Research Center (NIMed), Faculty of Physics, Department of Applied and Theoretical Physics Pontifícia Universidade Católica do Rio Grande do Sul PUCRS.

Professional Experience

March 2013 – November 2016 Medical Physicist: *Supervisor of Radiation Protection in* Nuclear Medicine and PET/CT, QA, QC, and other activities involving SPECT, SPECT/CT and PET/CT in clinical practice. Santa Casa de Misericórdia de Porto Alegre – Santa Rita Hospital Porto Alegre, Brazil

March 2012 – November 2016 Radiologic Technology Instructor: Lectures on Physics in Nuclear Medicine and Industrial Radiology. Saint Pastous Radiologic Technical School – Porto Alegre, Brazil

September 2011 – January 2013 Medical Physicist: *Supervisor of Radiation Protection in* Nuclear Medicine, Quality Control SPECT and PET/CT activities. NUCLEARMED Ltda. Porto Alegre, Brazil

August 2011 – September 2012 Technical Instructor: Lectures on Radiation Protection, Biological Effects, Physics and Math. Universitário Technical School – Porto Alegre, Brazil

August 2010 – December 2010 Visiting Technical Instructor: Lecture of Nuclear Medicine at ULBRA University - Technical School. Canoas, Brazil

January 2008 – December 2011 Medical Physicist: *Supervisor of Radiation Protection in Nuclear Medicine* – Quality Control of SPECT systems – UNITOM Clinic– Cascavel – PR/Brazil

March 2008 – June 2008 Medical Physicist: *Supervisor of Radiation Protection* in Nuclear Medicine. Hospital de Clínicas de Porto Alegre (HCPA). Porto Alegre, Brazil

March 2007 – July 2007 Laboratory Tutor: Lecture of SPECT Systems, assisting Prof. Dr. Ana Maria Marques of Physics Faculty, PUCRS. Porto Alegre, Brazil

March 2004 – September 2005 Intern: Administrative Support in Diagnosis and Treatment – Nuclear Medicine Facility at Hospital de Clínicas de Porto Alegre (HCPA). Porto Alegre, Brazil

Professional Appointments:

Cláudia Régio Brambilla is a PhD Researcher at Institute of Neuroscience and Medicine, Medical Imaging Physics, (INM – 4), Forschungszentrum Juelich GmbH, Jülich, Germany.

DAAD Scholarship. Research Grants - Doctoral Programmes in Germany, 2017/18.

Publications from the last 3 years:

Brambilla, Cláudia and Hoff, Gabriela. *Impact Evaluation of the Geometry in the Solid Radioactive Waste Exposure Rate Measurements in Nuclear Medicine.* RBEB - Brazilian Journal of Biomedical Engeneering, Oct. 2014, v.30 (4), p.330-340.

Gabriela Hoff
Dipartimento di Fisica, Università di Cagliari, Monserrato, Italy

Education: PhD on Bionuclear Science

Research and Professional Experience:

Research Experience

October 2016 – December 2016 – Post-Doc at Università degli Studi di Sassari (UNISS, Itália) in the project Validation of xrmc Monte Carlo Code for Dosimetry in Mammography. Grant to project protocol number 2446 applied as Visitor Professor, from Univertità degli Studi di Sassari.

September 2014 – October 2015- Post-Doc at Technologic Federal University of Paraná (Universidade Tecnológica Federal do Paraná – UTFPR) in the project Tomography using protons beam: building a cluster. Grant to project number 161/2012 applied as Visitor Professor, from Brazilian Ministry of Education - Higher Education Personnel raining Coordination (Coordenação de Aperfeiçoamento de Pessoal de Nível Superior - CAPES).

March 2014 – August 2014 - Adviser of the student Nathan Willig Lima in the project Implementation and validation of a methodology to solve the Bolztmann transport equation using the Klein-Nishina model to Compton Scattering. Pontifical Catholic University of Rio Grande do ul (Pontifícia Universidade Católica do Rio Grande do Sul – PUCRS) Physics Faculty – Medical Physics Course. Undergraduation scholarship to the student from Pontifical Catholic University of Rio Grande do Sul (Pontifícia Universidade Católica do Rio Grande do Sul – PUCRS).

May 2011 – September 2012 - Post-Doc at National Institute of Nuclear Physics (Istituto Nazionale di Fisica Nucleare – INFN), Italy, in the project Comparative study and electrons transportation modeling using Monte Carlo method. Grant identification BEX 6460/10-0, from Brazilian Ministry of Education - Higher Education Personnel Training

Coordination (Coordenação de Aperfeiçoamento de Pessoal de Nível Superior - CAPES).

2008 – 2010 - Adviser of the students Carla Alves Bork da Silva and Jefferson Santana Martins in the project Development of operational standard procedure and management of quality control data to medical and odontological x- radiation. Pontifical Catholic University of Rio Grande do Sul (Pontifícia Universidade Católica do Rio Grande do Sul – PUCRS) Physics Faculty – Medical Physics Course. Undergraduation scholarship to the student from Pontifical Catholic University of Rio Grande do Sul (Pontifícia Universidade Católica do Rio Grande do Sul – PUCRS).

2009 – 2011 - Researcher in the project Evaluation of Isoexposure Curves generated by mobile x-ray equipments. Pontifical Catholic University of Rio Grande do Sul (Pontifícia Universidade Católica do Rio Grande do Sul – PUCRS) Physics Faculty – Medical Physics Course.

2009 – 2010 - Adviser of the student Sandro Fernandes Firmino in the project Transmission curves determination and study of the influence of the material composition, commonly used as shielding, in diagnostic radiology. Pontifical Catholic University of Rio Grande do Sul (Pontifícia Universidade Católica do Rio Grande do Sul – Physics Faculty – Medical Physics Course. Undergraduation scholarship to the student from Pontifical Catholic University of Rio Grande do Sul (Pontifícia Universidade Católica do Rio Grande do Sul – PUCRS).

2007 – 2008 - Adviser of the student Pery Vidal in the project External dosimetry in mammography, a realistic study using computation simulation. Pontifical Catholic University of Rio Grande do Sul (Pontifícia Universidade Católica do Rio Grande do Sul – PUCRS) Physics Faculty – Medical Physics Course. Undergraduation scholarship to the student from Pontifical Catholic University of Rio Grande do Sul (Pontifícia Universidade Católica do Rio Grande do Sul – PUCRS).

January 2001 – July 2001 - Researcher invited to develop projects in Diagnostic Radiology area (computational and experimental dosimetry) Hospital of Clinics of Porto Alegre (Hospital de Clínicas de Porto Alegre – HCPA), Brazil.

1997 – 1998 - Training in Medical Physics and development of Master dissertation practical activities. University of Alabama at Birmingham, UAB, United States. Fellowship under the Prof. Dr. Gary T. Barnes at University of Alabama at Birmingham Hospital.

Among other research activities in 13 year as professor of PUCRS University I worked, as adviser, in 21 monographs of conclusion of Course and co-adviser in 4 master dissertations in different knowledge areas and other collaboration in research.

Professional Experience

June 2001 – July 2014 - Associate Professor (after 2005) Pontifical Catholic University of Rio Grande do Sul (Pontifícia Universidade Católica do Rio Grande do Sul – PUCRS) Physics Faculty – Medical Physics Course. Brasil. The teaching activities, in this period were 14-20 hours per week, in 40 hours of work per week. Other activities I performed were: Coordination of Medical Physics Course (Graduation) and participation of Ethics Committee in Research of the University and collaboration with internal events (conferences and symposiums) among other activities.

Publications From The Last 3 Years

Accepted for publication: Hoff, G; Filipov, D; Denyak, V.; Schelin, H. R.; Paschuk, S. A. *Monte Carlo Simulation used to Calculate Energy Correction Factor for Thermoluminescent Dosimeters used by Occupational Workers on Pediatric Exams*. Radiation Protection Dosimetry. 2017.

Hoff, G; Denyak, V; Schelin, H. R.; Paschuk, S. *Validation of Geant4 on Proton Transportation for Thick Absorbers: study case based on Tschalär experimental data.* IEEE Transactions on Nuclear Science, v. 6(2), p. 745 – 771, Feb. 2017.

Han, M. C.; Kim, S. H.; Pia, M. G.; Basaglia, T.; Batic, M.; Hoff, G; Kim, C. H.; Saracco, P.. *Validation of Cross Sections for Monte Carlo Simulation of the Photoelectric Effect.* IEEE Transactions on Nuclear Science, v. 63, p. 1117-1146, 2016.

Lima, N.; Hoff, G. *Impacto da pureza dos filtros de alumínio no valor de Camada Semi-Redutora em radiologia convencional e mamografia.* Brazilian Journal of Radiation Sciences, v. 4, p. 1-13, 2016.

Basaglia, T.; Cheol Han, Min; Hoff, G.; Kim, C. H.; Kim, S. H.; Pia, M. Grazia; Saracco, P. *Quantitative Test of the Evolution of Geant4 Electron Backscattering Simulation.* IEEE Transactions on Nuclear Science, v. PP, p. 1-1, 2016.

Kim, S. H.; Pia, M. Grazia; Basaglia, T.; Han, M. C.; Hoff, G.; Kim, C. H.; Saracco, P.. *Investigation of Geant4 Simulation of Electron Backscattering.* IEEE Transactions on Nuclear Science, v. 62 (4). p. 1805-1812. 2015.

Kim, S. H.; Pia, M. G.; Basaglia, T.; Han, M. C.; Hoff, G; Kim, C. H.; Saracco, P. *Validation Test of Geant4 Simulation of Electron Backscattering.* IEEE Transactions on Nuclear Science, v. PP, p. 1-29, 2015.

Brambilla, C. R.; Hoff, G. *Impact evaluation of the geometry on measurements of solid radioactive waste exposure rates in nuclear medicine.* Brazilian Journal of Biomedical Engineering, v. 30, p. 330-340, 2014.

Hoff, G.; Costa, P. R. *A comparative study for different shielding material composition and beam geometry applied to PET facilities: simulated transmission curves.* Brazilian Journal of Biomedical Engineering. v. 29(1), p. 86-96, 2013.

Batic, M.; Hoff, G; Pia, M. G.; Saracco, P.; Weidenspointner G. *Validation of Geant4 Simulation of Electron Energy Deposition.* IEEE Transactions on Nuclear Science, v. 60, p. 2934-2957, 2013.

In: Radioactive Wastes and Exposure   ISBN: 978-1-53612-213-8
Editor: Austin D. Russell   © 2017 Nova Science Publishers, Inc.

Chapter 3

THE CALCULATION OF DOSIMETRY IN SMALL ANIMALS COMBINING THE FDTD METHOD AND EXPERIMENTAL MEASURES: APPLICATION FOR THREE RADIATION SYSTEMS

Elena López-Martín[1,*], MD, PhD,
Aarón A. Salas-Sánchez[2], Alberto López-Furelos[1],
Francisco J. Jorge-Barreiro[1], MD, PhD,
Eduardo Moreno-Piquero, PhD[2]
and Francisco J. Ares-Pena, PhD[2]

[1]Department of Morphological Sciences, Faculty of Medicine,
University of Santiago de Compostela, North Campus,
Santiago de Compostela, Spain
[2]Department of Applied Physics, Faculty of Physics,
University of Santiago de Compostela, Life Campus,
Santiago de Compostela, Spain

---

* Corresponding Author Email: melena.lopez.martin@usc.es.

Abstract

This chapter presents an assessment of a mixed system for calculating electromagnetic dosimetry that combined experimental radiation parameters and Finite Differences in Temporal Domain (FDTD) simulations based on numerical phantom rats. Using a simple formula that accounts for the different radiation exposure variables, it was possible to determine the specific absorption rate (SAR) in tissues of animals exposed to non-ionizing radiation in three experimental systems (standing wave cavity, traveling wave cavity and multifrequency). To estimate the SAR values for these three experimental systems, it was necessary to combine the measured values of the power absorbed by the animal (in the standing wave cavity) or the value of the $|E|$ field (a Gigahertz Transverse Electromagnetic (GTEM) chamber was used for traveling wave, single frequency, and multifrequency experiments) with FDTD numerical computations. The three experimental systems were analyzed and different biological models were compared. The authors also establish a discussion on the methodological evolution of dosimetry calculation and the biological results obtained in these three experimental radiation systems.

Keywords: dosimetry, FDTD, animals, SAR, GTEM, standing wave cavity, travelling wave cavity, multifrequency

1. Introduction

The determination of precise dosimetry is essential to research on harmful biological effects associated with radiofrequency (RF). Dosimetry is studied in biological models in an attempt to obtain specific absorption rate (SAR) values that can be replicated or confirmed for the entire body or particular body areas in living beings.

For a few years now, high resolution digital anatomical models that include permittivity values in human and animal tissues have made it possible to use the Finite Differences in Temporal Domain (FDTD) technique for evaluating the amount of energy absorbed. From the initial description of the FDTD calculation to localized prediction based on standardized whole-body SAR (W/kg) (Kunz et al., 1993), numerical precision has advanced significantly in commercial software simulation processes. The digital anatomical model of the Sprague Dawley rat (Schmid & Partner Engineering, 2009) used in research is based on magnetic resonance (MRI) data from different tissues and the numerical calculation to predict SAR values. However, with this methodology alone, the radiation model used for dosimetric calculation cannot be validated experimentally.

In this paper, we describe the SAR calculation for animals exposed to RF in three different experimental systems. A simple formula that combines experimental measurements of the power absorbed by the animal with values obtained from FDTD simulation makes it possible to determine the specific absorption rate with greater precision.

2. MATERIALS AND METHODS

2.1. Description of Experimental Radiation Systems

2.1.1. Experimental System I: Standing Wave Cavity

This experimental system consists of a metal cavity, in which the animal is placed inside a methacrylate holder in the area of maximum radiation. The dimensions are large enough to minimize animal stress. A transmitting antenna (TA), a receiving antenna (RA) and a video camera were also placed inside the cavity.

The frequency, amplitude and modulation of the signal were established by a generator, which was connected to an amplifier. The amplified signal is transmitted through a directional coupler to the transmitting antenna of the cavity, where the radiation takes place. A pure 900 MHz sinusoidal signal was used for the experiments.

A receiving antenna (RA) connected to a spectrum analyzer was installed to monitor and check field stability while ensuring the absence of spurious signals.

A subsystem composed of directional couplers, sensors, and power meters made it possible to calculate the power absorbed by the animal as the difference in power absorbed by the system with or without the rat inside the cavity.

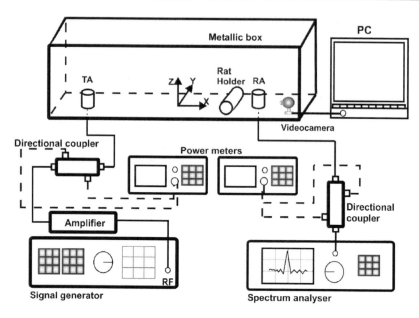

Figure 1. Diagram of the experimental set-up. Agilent signal generator, spectrum analyser and power meters (Models E4438C, E4407B and E4418B, respectively), Aethercomm linear power amplifier, and Narda directional couplers (Model 3282B-30). TA, transmitting antenna; RA, receiving antenna. The origin of the coordinates is located at the centre of the chamber floor.

The region where the animal is placed for radiation was previously determined by commercial FDTD software (SEMCAD), with only the transmitting antenna (monopole $\lambda/4$) inside the box and considering the region of radiation to be limited by a perfect electrical conductor (PEC), in addition to the metallic structure. A calculation of the distribution of the |E| field created by the transmission antenna was also necessary in order to determine the optimal positions for both the rat and the receiving antenna.

2.1.2. Experimental System II: Traveling Wave Cavity

In this experimental system, the vector signal generator (VSG) signal feeds the amplifier a pure 2.450 MHz sinusoidal signal adjusted to the power required for radiation. The amplifier output is connected to the directional coupler (DC) and passes directly into the GTEM radiation chamber. The rat (R) is immobilized in a methacrylate holder (RH) and irradiated in the zone of maximum field uniformity, with the left front leg of the animal receiving maximum radiation. The DC can measure incident and reflected power, monitoring the value of the first through the spectrum analyzer and obtaining the value of the second with the power meter.

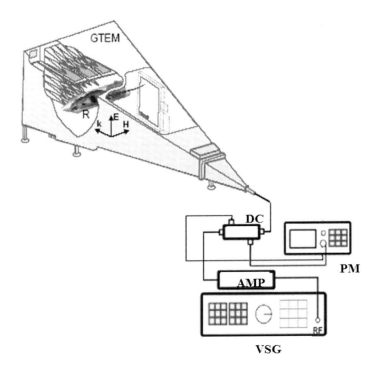

Figure 2. Illustration of the experimental set-up. GTEM cell (Schaffner 250 irradiation chamber); VSG: Vector Signal Generator (Agilent E4438C: 250 kHz-4 GHz); AMP: Amplifier (Aethercomm 0.8-3.2-10); DC: Directional Coupler (NARDA 3282B-30: 800-4000 MHz); PM: Power Meter (Agilent E4418B); RH: Rat Holder.

An isotropic probe is used to measure the field and determine the peak value. This measurement is performed before placing the rat into the chamber, using the values of the desired input signal. In this way, we can define the behavior of the camera in the measurement area.

2.1.3. Experimental System III: Multifrequency System

This system consists of two vector signal generators (VSG1 and VSG2), each of which generates a pure sinusoidal signal of 900 MHz and 2450 MHz, respectively, at the required power during radiation. The output from both generators is connected to a signal mixer (SM) that passes the combined signal to the amplifier (AMP).

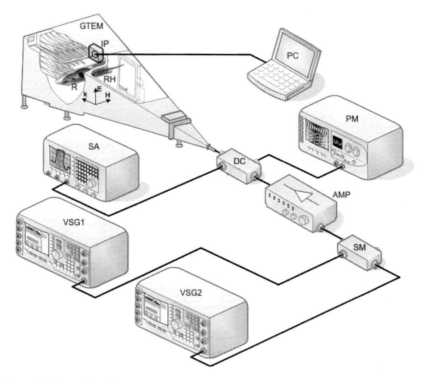

Figure 3. Schematic of the experimental set-up. GTEM, Schaffner 250 GTEM chamber; VSG1, Agilent E8267D vector signal generator (250 KHz–20 GHz) operating at 2.45 GHz; VSG2, Agilent E4438C vector signal generator (250 KHz–4 GHz) operating at 900 MHz; AMP, research amplifier 15S1G3 (0.8–3 GHz); DC, NARDA 3282B-30 directional coupler (800–4000 MHz); SA, Agilent E4407B spectrum analyzer (9 KHz–26.5 GHz); PM, Agilent E4418B power meter; SM, Agilent 11636a signal mixer; RH, rat holder; IP, EF Cube isotropic probe; R, rat.

Once amplified, the signal is sent to the directional coupler (DC) and passes directly into the GTEM radiation chamber where the rat (R), immobilized the methacrylate rat holder (RH) is positioned in the zone of maximum field uniformity. The Power Meter (PM) measures incident power values from the DC in order to establish the desired input power for the system. Reflected power is also measured and monitored through the spectrum analyzer (SA), to check the spectral purity of the sinusoidal signal used in this experiment.

The field impinges on R in the direction k, with the vectors E and H positioned perpendicular and parallel to the principal axis of R, respectively. Consequently, the left zone of R receives the maximum field amplitude. The Isotropic Probe (IP) measures the field and provides its peak value. This measurement is performed using the desired values in the input signals, without placing the rat inside the chamber, in order to specify the behavior of the camera in the measurement area. This value will be used later to achieve a more objective simulation of the GTEM chamber, using two plane wave fronts that will reproduce the data obtained with the probe.

2.2. Calculation of SAR

The SAR values obtained from the animals studied in each of the three experimental systems were estimated using commercial FDTD software based on SEMCAD X (Schmid & Partner Engineering AG, 2009). The phantom animal used was a 198.3g Sprague-Dawley numerical rat model (Schmid & Partner Engineering AG, 2009), assembled in 1.15mm sections (obtained by magnetic resonance imaging) and composed of 60 different tissues.

The numerical model was obtained by simulating a standing wave in the radiation cavity of the first experimental system and one or more plane waves used indistinctly in the GTEM chamber of the second and third experimental systems. The signal affected the left side of the animal, with the magnetic field H parallel to its main axis. The simulations (executed on a desktop PC with a 3.20 GHz Intel Core i7 processor, 24 GB RAM, and an Nvidia Tesla C-1060 graphics card adapted for numerical calculation) were performed at 900 MHz and 2.45 GHz or by combining both frequencies. It took 400 periods to reach steady state in the simulation.

For Experimental Systems II and III, the SAR estimates were obtained by applying a correction factor to the simulation values that was proportional to the weight of the numerical rat with respect to the weights of the animals used during the experimentation:

$$SAR_E = SAR_S \times W_S / W_E \qquad (1)$$

where SAR_E is the experimental SAR estimate, SAR_S is the SAR value obtained during the simulation, $W_S = 198.3$ (g) is the weight of the numerical model, and W_E (g) is the weight of the animal that was used in the experiment. In Experimental System I, it was necessary to multiply the second term of Equation (1) by P_E/P_S, where P_E is the power absorbed by the animal and P_S is the power absorbed by the numerical rat, determined by simulation.

3. RESULTS

3.1. Experimental System I: Application in a Subconvulsive Model of Picrotoxin in Rats

In this study, we observed the effects after two hours of modulated radiation at 900MHz GSM in a seizure-prone rats model that that simulated injection of a subconvulsive dose of picrotoxin (antagonist to GABA-A receptors) (Nutt et al.,1988). The effects of radiation were assessed by video-recorded observation (supported in some cases by electroencephalography) of rats who suffered seizures, along with postmortem studies involving immunochemical testing for c-Fos-positive brain areas. This sensitivity marker for neuronal activation (Morgan et al., 1999) indicates the occurrence of electric discharge in the brain (Willoughby et al., 1995). The experimental design and set up of the standing wave cavity included a field characterization assay with the rat located at the point of maximum radiation in the metal box (see figure 4).

As discussed in the previous section, SAR_E values were obtained from: the simulated SAR product, the ratio between absorbed power for each real rat and absorbed power for the numerical rat, and the ratio between the actual weight of each animal and the simulated weight. To obtain sub-thermal SAR values, 0.25-1W power was used (see Table 1).

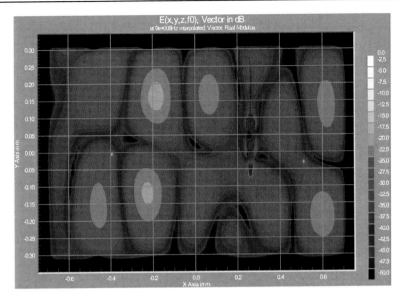

Figure 4. The figure represents SEMCAD-simulated distribution of |E| on a horizontal plane at the level of the phantom's head, with the phantom rat, the transmitting antenna, and the receiving antenna inside the chamber.

Table 1. SAR$_E$ values calculated from the SARs and the power absorbed by the animals in the standing wave cavity

Mean absorbed power (mW)	Weight (g)	Estimated SAR (mW/kg)			
		Mean in brain (mW/kg)	Peak average in 1 g of brain (mW/kg)	Whole body mean SAR (mW/kg)	Peak average in in 1 g of whole body (mW/kg)
56.77	209.95	0.24	0.27	0.24	1.29
192.67	225	1.32	1.49	0.74	4.11

The neurological results for the animals radiated in this cavity were published in (López-Martín et al., 2006, 2009; Carballo-Quintá et al., 2011).

3.2. Experimental System II: Application in a Diathermy Model

In this study, HSP 70 and 90 heat shock proteins and glucocorticoid receptors were used to determine cellular stress levels. These were analyzed by performing ELISA and immunohistochemistry tests, and by H&E staining to reveal histological changes in the rat thymus after exposure to 2.45 GHz radiofrequency in a GTEM chamber using a rat diathermy model.

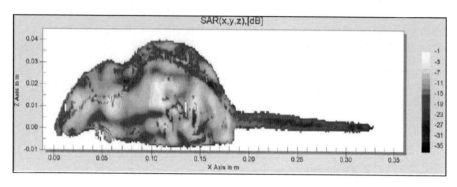

Figure 5. Distribution of local SARs in the phantom rat 'exposed' to 2.45 GHz at 3 W, in the plane X = 0.33 m.

The effects of radiation on thymus tissue, using this experimental system, were published in (Misa-Agustiño et al., 2015).

Table 2. SAR values in thymus and body of experimental rats, calculated from power (P) and electric field (E). Results are expressed as mean ± standard error of the mean (M±SEM), obtained from mean SAR and average peak SAR in the animals. Values were compared by one-way ANOVA for the power applied during radiation (0, 1.5, 3, 12 W), followed by the Holm–Sidak test for multiple comparisons

	FDTD-calculated specific absorption rate: experimental measurements			
	Mean SAR in thymus (W/kg)	Peak SAR in 1 g of thymus	Mean SAR in body (W/kg)	Peak SAR in 1 g of body
P = 1.5 W E = 28.48 V/m	0.046 ± 1.10^{-3}	0.041 ± 2.10^{-3}	0.0169 ± 7.10^{-4}	0.089 ± 9.10^{-3}
P = 3 W E = 40.28 V/m	0.104 ± 5.10^{-3}	0.076 ± 4.10^{-3}	0.0364 ± 19.10^{-3}	0.180 ± 9.10^{-3}
P = 12 W E = 80.56 V/m	0.482 ± 12.10^{-3}	0.340 ± 10.10^{-3}	0.161 ± 4.10^{-3}	0.795 ± 2.10^{-3}

The Calculation of Dosimetry in Small Animals ... 137

The SAR_E values for each rat were obtained by applying a correction factor – the ratio between the weight of the simulated rat and the weight of the experimental animal – to the SAR values obtained through numerical simulation (Expression 1). Table 1 shows the mean SAR or peak SAR values for 1g of the thymus and/or whole body of the rat, expressed as SAR±SEM, indicating significant differences (p <0.001) between SAR values obtained from different power levels. Note that the increase of the mean and peak SAR in each group is directly proportional to the incident power. The local distribution of SARs in the rat tissues is shown in Figure 5.

3.3. Experimental System III: Application in a Male Rat Model

In this study, rats were exposed to non-thermal levels of radiation at 2450 MHz, 900 MHz, or both simultaneously, in a GTEM chamber. Peak SAR was estimated (specific absorption rate in 1g of tissue) in various tissues using the numerical rat model and FDTD software. To calculate simulated SAR for simultaneous radiation with 900 and 2450 MHz signals, the mean SAR values were obtained for 900 and 2450 MHz separately, then expression (1) was used to obtain SAR_E values (see Table 3 and Figure 6). Studies were performed on apoptotic tissues using DAPI and H&E (López-Furelos et al., 2012).

Table 3. Experimental conditions, weights, and estimated 1g average SARs for whole body peak and cerebral hemispheres in the experimental groups subjected to RF radiation

Group	f (MHz)	P_{TR} (W)	E_m (V/m)	SAR_E (W/kg) (Whole body)	SAR_E (W/kg) Cerebral Hemispheres
I	900	2	47.5	0.171	0.090
II	2450	2	40.2	0.068	0.074
III	900 2450	1+ 1	34.4	0.132	0.10

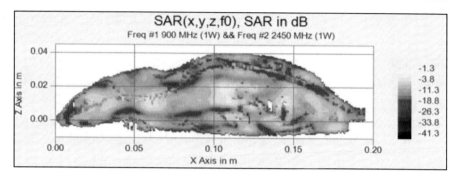

Figure 6. Distribution of 1g average SAR in horizontal sections of the numerical rat phantom when radiated simultaneously with 900MHz (PTR = 1W) and 2450MHz (PTR = 1W). SAR is expressed relative to absorption of the entire local power density in the absence of the rat. Scale in standard dB. Vertical cut (main plane y = constant) coincident with the axis of the rat.

4. Discussion

In this work, we described and analyzed a tool for estimating SAR values in the laboratory for rats exposed to radio frequency fields in electromagnetic experiment chambers. The combined calculation of numerical simulations with a phantom rat using FDTD and experimental measurement of the electromagnetic parameters in situ made it possible to approximate the biological index of the energy absorbed by the animals, which is required for studies of biological effects. Experimental validation of the combined dosimetry calculation involved the application of this methodology in three different radiation systems. In Experimental System I, electromagnetic tests were done in a standing wave cavity in order to study a pre-convulsive rat model (López-Martin et al., 2006, 2008, 2009; Carballo-Quintá et al., 2011) that made it possible to measure the power absorbed by the animals (López-Martín et al., 2008) and estimate the dosimetry calculation. The sub-thermal SAR values obtained indicated that the biological effects observed were not due to increased temperature or stress from immobilization, as other authors have indicated (Finne et al., 2005). Similarly, using a diathermy model in Experimental System II, SAR data for the brain, thyroid and thymus of animals led us to conclude that apart from the effects caused by increased tissue temperature, the interaction of electromagnetic fields with tissues caused a wide variety of acute transitory changes that sometimes persisted in the tissue physiology and morphology (Jorge-Mora et al., 2010, 2011; Misa-Agustiño et al., 2012, 2015). The third experimental system made it possible to observe biological effects after the interaction of more than one radiofrequency. The results offered no biological evidence of the cumulative effects that might be expected if the energy absorbed in the combined frequencies case corresponded to the addition of the energy of both frequencies. Thus, it is necessary to look for other mechanisms to explain tissue interactions resulting from combined radiofrequencies in mammals. Our hypotheses were based on biological experimental evidence and supported by dosimetry indexes obtained by combining numerical simulation with radiation experiments. Dosimetry calculation has been a primary tool for analyzing the biological results in order to interpret the mechanisms that explain the findings. However, though SAR estimates help to interpret the results of radiation exposure in live animal tissues, it is necessary to evaluate this under controlled experimental conditions, inside the radiation chamber. The electromagnetic parameters used in the experiment – time, frequency, power, energy absorbed by the animals, temperature – are highly relevant and should accompany the dosimetry. There is substantial criticism for the use of numerical models, such as the FDTD method, which simulate the spatial distribution of the radiation energy for the SAR calculation. Most of the criticism is aimed at the possible errors that arise from the use of overly simplified numerical models, which do not account for the variations in physical parameters at the cellular level that occur in live tissue (Panagopoulos et al., 2013). While these are valid points, it is important to remember than a numerical model is always an approximation of reality. Other options for estimating dosimetry in tissue would involve the insertion of micro-antennas in the tissue to detect internal fields or the use of thermometer probes or a thermistor to detect tiny changes in tissue temperature (Gong et al., 2016; Stauffer et al., 2013). However, these would also introduce errors due to simplification or from altering the features of the tissue under study. Perhaps the most precise way of estimating SAR would be through a combined methodology involving numerical models and measurements of internal fields and temperature in the tissues. This estimate might be more accurate, but would undoubtedly be more complex, sophisticated and costly. The experience acquired in our laboratory over many years of dosimetry calculation using numerical rat models (López-Martin et al., 2006) or through the detection of changes in rectal temperature (López-Vila et al., 2015) has allowed us to see the limitations of both methods. However, one important reason for obtaining the power absorbed by the animal and the SAR after exposure to non-ionizing radiation is to ensure that the energy absorbed by the animals is below thermal levels, by controlling electromagnetic parameters in the test chamber (López-Martin et al., 2008, Jorge-Mora et al., 2010), environmental temperature and the rectal temperature of the animals (Jorge-Mora et al., 2011; Misa-Agustiño et al., 2012). In this way, sub-thermal SAR values can be obtained for the tissues, organs or whole body to complement the biological tissue studies. To our surprise, many of the dosimetry values obtained in our experiments did not allow us to establish a quantitative relation between dosage and biological response (Misa-Agustiño et al., 2012), indicating that in many cases there was no linear relation between electromagnetic exposure and biological effects (López-Furelos et al., 2016). The specific absorption rate (SAR) cannot be considered in isolation; the characteristics of the cavity where the radiation was applied and the biological experimental model used must also be taken into account. Thus, we applied sub-thermal absorption levels (SAR<4W/kg) of non-ionizing radiation to rats in different experimental contexts: in a standing wave cavity with a subconvulsive picrotoxin model (Carballo-Quintás et al., 2011) and a traveling wave cavity (Jorge-Mora et al., 2010) with a diathermy model, we obtained very different biological effects. In contrast, we used the same experimental cavity, a standing wave, and the same biological model to expose the animals to radiofrequencies with and without modulation of sub-thermal SAR, and obtained different results (López-Martín et al., 2009). In more complex experimental radiation models, where we looked at the interaction of multiple radiofrequencies, we saw that SAR constituted a contradictory biological index for explaining the biological effects of cellular stress (López-Furelos et al., 2016). Again, when the complexity of

The Calculation of Dosimetry in Small Animals ... 141

the experimental model increased, we found no linearity between exposure to electromagnetic fields and the biological effects observed.

Dosimetry calculation (SAR) is a tool that allows researchers to establish a range in the level of radiation. It gives information about the energy absorbed by the tissues, but by itself does not give information about biological effects. Methodology that combines digital numerical models based on FDTD with experimental data about RF electromagnetic parameters is non-invasive and the relevant calculations can be replicated and confirmed. In spite of its limitations, and the need to include a range of error (Kuster et al., 2006), it is a necessary index for interpreting biological effects within the electromagnetic model studied in the laboratory.

CONCLUSION

The experimental validation of this combined dosimetry calculation involved the application of this methodology in three different radiation systems:

1) In an electromagnetic test cavity with a stationary wave, in order to study a pre-convulsive rat model, which facilitated the measurement of the power absorbed in the body and brain of the animals, and estimation of SAR by combining it with the SARs value.

2) Experimental measurement of the value of the $|E|$ field in the GTEM chamber, which enabled the calculation of simulated SAR and the subsequent value of SAR_E in the brain, thyroid and thymus of animals in a diathermy model.

3) Combining experimentally determined measurements of absorbed power and the $|E|$ field with numeric calculations using FDTD made it possible to estimate SAR for one or two frequencies/powers in the three experimental systems.

The dosimetry values we obtained from our experiments did not support the establishment of a quantitative relationship between dose and biological response or the subsequent establishment of a linear relationship between electromagnetic exposure and biological effect. SAR makes it possible to establish a range of radiation levels to complement the study of the biological effects observed in each model.

ACKNOWLEDGMENTS

The authors are grateful to the Ministerio de Economía y Competitividad for funding awarded through project TEC-2011–24441. We also greatly appreciate the assistance provided by Jose Carlos Santos and Rafael Fuentes.

REFERENCES

Carballo-Quintá, M., Martínez-Silva, I., Cadarso-Suárez,C., Alvarez-Folgueiras, M., Ares-Pena, F. J., López-Martín, E. (2011) A study of neurotoxic biomarkers c-fos and GFAP after acute exposure to GSM radiation at 900MHz in the picrotoxin model of rat brains. *Neurotoxicology*, 32, 478-494.

Finnie, J. W. (2005) Expression of the immediate early gene, c-fos, in mouse brain after acute global.

Gong, Y., Capstick, M., Tillmann, T., Dasenbrock, C., Samaras, T., and Kuster, N. (2016) Desktop Exposure System and Dosimetry for Small Scale In Vivo Radiofrequency Exposure Experiments. *Bioelectromagnetics* 37:49-61.

Jorge-Mora, M. T., Alvarez-Folgueiras, M., Leiro, J., Jorge-Barreiro, F. J., Ares-Pena, F.J., Lopez-Martín, E. (2010) Exposure to 2.45 GHz microwave radiation provokes cerebral changes in induction of HSP90 α/β heat shock protein in rat. Progress. In Electromagnetics Research, *PIER* 100,351-379.

Jorge-Mora, M. T., Misa-Agustiño, M. J., Rodriguez-Gonzalez, J. A., Jorge-Barreiro, F. J., Ares-Pena, F. J., Lopez-Martín, E. (2011) The effects of single and repeated exposure to 2.45 GHz radiofrequency fields on c-fos protein expression in the paraventricular nucleus of rat hypothalamus. *Neurochem. Res.*, 36:2322–2332.

Kunz, K. S., Luebbers R. J. (1993). The Finite Difference Time Domain Method for Electromagnetics, CRC Press, Inc; Boca Raton, Florida.

Kuster, N., Torres, V. B., Nikoloski, N., Frauscher, M., Kainz W. (2006) Methodology of detailed dosimetry and treatment of uncertainty and variations for in vivo studies. *Bioelectromagnetics;* 27(5):378-91.

López-Furelos, A., Leiro, J. M., Rodriguez-Gonzalez, J. A., Miñana-Maiques, M. M., Ares-Pena, F. J., López-Martín, E. (2012) An Experimental Multi-frequency System for Studying Dosimetry and Acute Effects on Cell and Nuclear Morphology in Rat Tissues. Progress in Electromagnetics Research, *PIER*. 129, 541-558. 8.

López-Furelos, A., Leiro, J. M., Salas-Sánchez, A. A., Ares-Pena, F. J., López-Martín, E. (2016) Evidence of cellular stress and caspase-3 resulting from a combined two-frequency signal in cerebrum and cerebellum of Sprague- Dawley rats. *Oncotarget*, 7, 40.

López-Martín E., Bregains, J., Relova-Quinteiro, J. L., Cadarso-Suárez, C., Jorge-Barreiro, F. J., Ares-Pena, F. J. (2009) The action of pulse-modulated GSM radiation increases regional changes in brain activity and c-fos expression in cortical and subcortical areas in a rat model of picrotoxin-induced seizure proneness. *J. Neurosci. Res.*, 87(6):1484–99.

López-Martín, E., Relova-Quinteiro, J. L., Gallego-Gómez, R., Peleteiro-Fernandez, M., Jorge-Barreiro, F. J., Ares-Pena, F. J (2006) GSM radiation triggers seizures and increases cerebral c-fos positivity in rats pretreated with subconvulsive doses of picrotoxin. *Neurosci. Lett.* 298: 139–44

López-Martín, E., Bregains, J. C., Jorge-Barreiro, F. J., Sebastian-Franco J. L., Moreno-Piquero, E., Ares-Pena, F. J. (2008) An experimental set up for measurement of the power absorved from 900 MHz GSM standing wave by small animals, illustrated by application to picrotoxin ttreated rats. Progress In *Electromagnetics Research, PIER* 87, 149–165.

López-Vila, L., Comparative analysis of specific absorption rate (SAR) and stress in small animals exposed to two radiation systems operating at 900MHz: Traveling wave and standing wave. (2015) These for MSc degree, USC.

Misa-Agustiño, M. J., Leiro, J. M., Jorge-Mora, M. T.,Rodríguez-Gonzalez, J. A., Jorge-Barreiro, F. J., Ares-Pena, F. J., and López-Martín, E. (2012). Electromagnetic fields at 2.45 GHz trigger changes in heat shock proteins 90 and 70 without altering apoptotic activity in rat thyroid gland. *Biology Open* 1, 831–838 doi: 10.1242 /bio.20121297.

Misa-Agustiño, M. J.,Leiro, J. M., Gomez-Amoza, J. L., Jorge-Mora, M. T., Jorge-Barreiro, F. J., Salas-Sánchez, A., Ares-Pena, F. J., López-Martín, E. (2015) EMF radiation at 2450 MHz triggers changes in the morphology and expression of heat shock proteins and glucocorticoid receptors in rat thymus. *Life Sciences* 127, 1–11.

Morgan, J. I., Curran, T. (1991) Stimulus-transcription coupling in the nervous system: involvement of the inducible proto-oncogenes fos and Ju *Annu. Rev. Neurosci.* 14, 421–451.

Nutt, D. J., Cowen, P. J., Batts, C. C., Grahame-Smith D. G.,Green, A. R. (1982) Repeated administration of subconvulsant doses of GABA antagonist drugs. I. Effect on seizure threshold (kindling), *Psychopharmacology* 76, 84–87.

Panagopoulos, D. J., Johanso, O., Carlo, G. L. (2013) Evaluation of specific absorption rate as dosimetric quantity for electromagnetic fields bioeffects. *Plos One.* 8, 6.

Schmid & Partner Engineering AG. (2009). Reference manual for the SEMCAD simulation platform for electromagnetic compatibility, antenna design and dosimetry. Available from http://www. semcad.com.

Stauffer, P. R., Rodriques, D. B., Salahi, S., Topsaka, E., Oliveir,a T. R., Prakash, A., D'Isidoro, F., Reudink, D., Snow, B. W., Maccarini, P. F. (2013).Stable Microwave Radiometry System for Long Term Monitoring of Deep Tissue Temperature. *Proc. SPIE Int. Soc. Opt. Eng.* 26; 8584.

Willoughby, J.O., Mackenzie, L., Medvedev, A., J. Hiscock, J. (1995) Distribution of fos-positive neurons in cortical and subcortical structures after picrotoxin-induced convulsions varies with seizure type, Brain Res. 683, 73–85.

In: Radioactive Wastes and Exposure ISBN: 978-1-53612-213-8
Editor: Austin D. Russell © 2017 Nova Science Publishers, Inc.

Chapter 4

EVOLUTION OF THE INTERNATIONAL LAW ON WASTES DISPOSAL IN GEOLOGICAL ENVIRONMENTS IN BRICS

Daniel Figueira de Barros and Daniel Marcos Bonotto[*]

[1]Centro Universitário Salesiano São Paulo (UNISAL), Americana,
São Paulo, Brasil
[2]Departamento de Petrologia e Metalogenia,
Universidade Estadual Paulista (UNESP), Rio Claro, São Paulo, Brasil

ABSTRACT

The world's concern over where to store radioactive waste gained prominence in 1978 at a technical committee meeting held in London coordinated by the International Atomic Energy Agency of the United Nations (IAEA), in order collect information about regulations and experiences from several attendee countries as well as they addressed regulatory approaches, thus there was a large discussion on the various aspects and issues involved in the subject. Primordially, at the time it was defined that repository systems were to include the burial of wastes in

---

[*] Corresponding Author Email: danielbonotto@yahoo.com.br.

either shallow or deep depths, also the disposition of radioactive wastes should take place in caves and in continental geological formations. At present, the final disposal of radioactive wastes in a safe way, in the light of the current, scientific and technological development, allows two main possible final destinations: the disposal in the environment or the confinement into the so-called final repositories. Confinement implies the definitive waste isolation inside a repository for long periods of time – from hundreds to thousands of years, depending on the half-life of the radionuclide. The purpose of this chapter is to describe and compare the legislation and legal provisions on waste disposal for medium, low and high levels of radiation among BRICS countries – Brazil, Russia Federation, India, People's Republic of China and South Africa. This chapter compares regulatory provisions, legislation, international agreements and competences of entities and/or international regulatory governmental bodies, showing BRIC'S status of law considering if their laws need improvement and reconsideration. Moreover, the chapter aims to report, in a short way, the historical evolution of international law on waste disposal in geological environment in BRICS by means of research on the international documentation and analysis based on bulletins from the Nuclear Energy Agency (NEA), a multinational and intergovernmental body part of the Organization for Economic Co-operation and Development (OECD) affiliated to the IAEA.

Keywords: environmental law, radiation, BRICS, nuclear law, radioactive waste disposal, geologic repository

INTRODUCTION

The civil nuclear industry, by means of nuclear power plants, became stronger after World War II, and it is very used in many countries. However, the global concern in seeking alternative energy sources to conventional ones (coal, oil, biodiesel and hydroelectric) is based on an attempt to decrease the emission of carbon dioxide (CO_2), as the world observes an increased energy demand and scarcity of fossil and water resources. Consequently, nations are discussing solutions to alleviate the drawbacks of global warming and its consequences.

Among the alternatives to large-scale power generation, nuclear might be an option for such worries towards the world economy and demand for energy with less CO_2 emission.

In this context, nuclear power plays multiple roles: 1) as thermal process and electricity sources which are economically efficient and environmentally advantageous; 2) as an industrial and intensive sector in knowledge and advanced technologies, capable of providing countless benefits to the economy as a whole; 3) as a focus of a revolutionary progress in science and technology in decades to come; and 4) as the focus of a change in cultural and educational paradigm, centered on the idea of man's participation in a nuclear universe (Tennenbaum, 2009).

Tennenbaum (2009) points out that current nuclear-electric-generation programs for a second electrification of nations require parallel efforts in order to reach technical improvements, and with respect to nuclear energy, the primary difficulty in generating energy is the radioactive waste produced by nuclear power plants. The problem is not the lack of *ad hoc* practical solutions for management, reprocessing and storage radioactive waste, in the medium term, because some solutions already exist and are incorporated into the routine operations of the nuclear industry.

For hundreds of years, the oceans have been used as a place to deposit and dump the waste from human activities, and in this regard, Low Level Waste (LLW) and Intermediate Level Waste (ILW) were dumped at more 50 locations in the northern part of the Atlantic and Pacific Oceans. However, it is important to say that High Level Waste (HLW), mainly the one coming from nuclear power plants, has not been dumped at sea (Calmet, 1989).

It is accepted that the disposal of radioactive waste is an activity that must be properly regulated by governments in order to achieve and maintain the necessary protection for mankind and environment from any potential danger from radioactivity. Then, regulatory provisions play a key role in the study, implementation and control of a national system for the proper disposal of waste and residues. Therefore, for HLW, a solution adopted by several countries was storage in geological repositories (IAEA, 1980).

For the present study, the literature was mainly used based on Nuclear Law Bulletin (NLB) publication, belonging to the Nuclear Energy Agency (NEA), entity affiliated with the International Atomic Energy Agency (IAEA) and the Organization for Economic Cooperation (OECD). It has published since 1968 legal regulations, agreements, legislative changes, legal case studies and other issues related to the legislation from several countries of the world with the scope of publicity for legislative comparison of nations as well as the standardization, which obviously will depend on country-specific circumstances (NLB, 1968).

Under this perspective, it has been compiled a brief comparison of laws from BRICS (Brazil, Russia, India, China and South Africa) concerning to its regulatory provisions for the radioactive waste disposal in geological repositories, mainly focusing on nuclear reactors' and HLW reprocessing plants'.

THE BRICS CONCEPTION

The BRICS conception was formulated by Goldman Sachs chief economist Jim O'Neil in a 2001 study entitled "Building Better Global Economic BRICs." The acronym has been established in economic-financial, business, academic and communication media, and back then in 2006, such concept gave rise to a grouping, incorporated into the foreign policy of Brazil, Russia, India and China. In 2011, at a III Summit of those countries, South Africa became part of the group, the "S" added to the acronym BRICS (GS, 2010; MRE, 2011).

The five BRICS countries in 2015 represented over 3.6 billion people, or half of the world population; all five members are in the top 25 of the world by population, and four are in the top 10 (IMF, 2013).

The BRICS members are developing or newly industrialized countries, being distinguished by their large, sometimes fast-growing economies and significant influence on regional affairs; all five are G-20 members (BBVA, 2012).

Since 2009, the BRICS nations have met annually at formal summits. According to Xinhuanet (2016), China will host the 9th BRICS summit in Xiamen on September 3rd-5th 2017. The term does not include countries such as South Korea, Mexico, and Turkey for which other acronyms and group associations were later created.

The BRICS' economic weight has certainly been considerable notwithstanding in 2003- 2007. The five countries' growth accounted for 65% of the worldwide GDP (gross domestic product). In purchasing power parity, BRICS' GDP exceeded the US's or the European Union's, amounting US$ 19 trillion or 25% of the world's purchasing power. In 2003, the group accounted for 9% of world's GDP, and in 2009 this figure increased to 14%. In 2010, the combined GDP of the five amounted US$ 11 trillion, or 18% of the world economy (MRE, 2011).

Overall, the BRICS expanded 4.6% in 2016, from an estimated growth of 3.9% in 2015. The World Bank expects BRICS growth to pick up to 5.3% in 2017 (EME, 2016).

As a group, BRICS have an informal character, there are no articles of incorporation, it neither works with a steady secretariat nor has funds to finance any of its activities. Ultimately, what sustains the mechanism is its members' political will. Despite the adversities that some countries have been passing, BRICS has had an institutionalization degree as the five countries intensify their interaction (Barros, 2012).

IAEA's Technical Committee on Regulatory Provisions and Underground Disposal Systems for Radioactive Waste

Hannes Olof Gösta Alfvén, Physics Nobel Prize 1970, stated two fundamental prerequisites for effective HLW management: (1) stable geological formations and (2) stability of human institutions over hundreds thousands (Balzani and Armaroli, 2011).

The world's concern over where to dispose radioactive material gained prominence in 1978 at a meeting of a technical committee held in London coordinated by IAEA. The meeting purpose was to collect information on regulations together with several countries, as well to address regulatory approaches, discussing various aspects and issues involved. Firstly, it was defined that repositories systems include either the waste burial at shallow depths or the waste disposal in caves and continental geological formations (IAEA, 1980).

This was one of the first legal records demonstrating the concern to regulate the new idea to dispose HLW in deep geological repositories. Representatives from 16 member states made presentations, as requested by the IAEA Secretariat, describing relevant national laws and regulations, as well as the role and responsibilities of government organizations and agencies involved such as licensing, inspection, and other related issues. This meeting gave rise to a technical report called "Regulatory Aspects of Underground Radioactive Disposal" which was published by the IAEA (IAEA, 1980).

Barros (2012) pointed out that, in the world, advances in relation to geological repositories have been made mainly by two countries: the US, with two geological repositories proposed for different radioactive materials (New Mexico and Nevada), and Sweden, by the company *Svensk Kärnbränslehantering AB (SKB)*, with the geological repository at *Östhammar* city.

It is important to highlight that the Joint Convention on the Safety of Spent Fuel Management and on the Safety of Radioactive Waste Management was adopted on 5 September 1997 by a Diplomatic Conference convened by the IAEA at its headquarters (Vienna) from 1 to 5 September 1997. The Joint Convention entered into force on 18 June 2001. In general terms, this Convention should apply to the safety of radioactive waste management when the radioactive waste results from civil applications, also to achieve and maintain a high level of safety worldwide in spent fuel and radioactive waste management, through the enhancement of national measures and international co-operation, including where appropriate, safety-related technical co-operation (IAEA, 1997). Currently, the contracting parties are Argentina, Australia, Austria, Belgium, Brazil, Canada, China, Czech Republic, Denmark, Estonia, EURATOM, Finland, France, Germany, Hungary, Ireland, Italy, Japan, Luxemburg, Poland, Romania, Russia, South Africa, Spain, Sweden, Switzerland, United Kingdom, United States (IAEA, 1997).

The existing laws of each BRICS country will be presented in this legislative approach, as well as the regulatory frameworks on the subject and its evolution to the present days. In the specific case of Brazil, the major information on the nuclear energy use in the country and the radioactive waste legislation have been already described by Barros and Bonotto (2015a, 2015b), thus, only the other BRICS countries will be here focused.

THE RUSSIAN FEDERATION (RF, RUSSIA)

In the former Union of Soviet Socialist Republics, the State ownership of the subsurface formed the public relations foundation for the use and conservation of mineral resources and enabled the planned and rational subsoil use. The legislation stated that the subsurface could be used for "(...) the construction and operation of underground facilities not being related to the minerals mining, including facilities for oil derivatives storage, gas and other materials, as well as disposal of harmful substances, industrial wastes and waste waters" (Pimenov et al., 1980).

Data on Russian legislation in the nuclear field became available for publication from the mid1980s, and from that time up to 1990 the country began to sign international treaties for nuclear accidents notification and technology exchanges with other countries (NLB, 1987a, 1987b, 1989b, 1991).

Only in 1996, it was published a NLB summary for the RF's nuclear legislation and its intention of modernizing the civil nuclear industry. But for that, the country needed to adjust its military and technical-scientific policies regarding the development and production of nuclear weapons, including safety measures. The country should also set goals to reduce and recycle nuclear weapons and seek solution for the disposal of its wastes.

On May 12, 1991, the USSR's President Mikhail Gorbachev signed the Act on social protection of citizens who have suffered Chernobyl accident damage. The Act contained topics on contaminated sites and decontamination and compensation, medicines provision, exams and disability pensions payment as well as compensation for citizens affected by that accident (NLB, 1991).

In November 1991, Guidance National Regulations applicable to Radiological and Nuclear Safety in the USSR Territory was issued, then, items were created to regulate the USSR's administrative structures, dismantling the RF's government reorganization and the extent of its jurisdiction. The Guidance also treated about sources and technological processes that used nuclear materials, nuclear power, radioactive sources in the territory and the reorganization and changes of national supervisory authorities' statutes (NLB, 1992).

Gosatomnadzor

The *Gosatomnadzor* (Federal Inspectorate for Nuclear and Radiation Safety) was established on December 31, 1991, as a regulatory body, which was responsible for preparation and improvement of national legislation on the energy production and use, nuclear materials, radioactive substances and nuclear field in general. This body still had the power to decide on the peaceful or military use of nuclear energy, and finally had to define the principles, criteria and safety standards, establishing the licensing and inspection system of nuclear activities (NLB, 1992).

MinAtom

In FR, nuclear field responsibilities were shared between the RF Ministry for Atomic Energy (MinAtom - created on January 29, 1992) being responsible for the national nuclear power program and the mentioned

Gosatomnadzor. It is important to say that until 1993 FR did not have a specific law regulating nuclear activities in the country, only two bills, one about the atomic energy use and another about a government policy on the management of radioactive waste, as well as apart texts on export, import, nuclear power plants and radioactive substances (NLB, 1994).

Until 1995, a law on radioactive waste management had not been signed by the President yet, despite an existing project (a specific Act) should treat exclusively on the nuclear energy use. In the RF, some legal provisions in the environmental area, human protection and health dated March 3, 1992, being considered as a non-precise basis about tailings disposal. Those rules prohibited the waste or radioactive materials disposal into the ocean floor, outer space and also the new 1995 Water Code, banned radioactive waste disposal in river basins (NLB, 2000).

In the RF, the law on the nuclear energy use separated federal agencies exercising control over the nuclear power use bodies from the state regulation system of safety. It is worth mentioning that the main regulatory body for the nuclear energy use was MinAtom, which performed technical and scientific research and organizational policies. That office also stated legal provisions and regulations on the use of nuclear power, being responsible for the control of nuclear materials and radioactive substances, thus, planning and implementing waste management programs (NLB, 2000).

Rosatom (POCATOM)

The state-owned company *Rosatom* Nuclear Energy State Corporation (currently known as the State Atomic Energy Corporation *ROSATOM)* took over the RF's nuclear industry in 2007, replacing the Federal Atomic Energy Agency (also known as *Rosatom* created on March 9, 2004) which was formed from the Minatom replacing the USSR Department of Nuclear Engineering and Industry (WNA, 2010a).

Atomprom

On 6 February 2007, the President signed the Federal Bill on the Management and Disposal of the Property and Shares of Organisations Operating Within the Country's Nuclear Energy Sector. The bill was adopted by the State Duma (Lower House of Parliament) on 19 January 2007 and approved by the Federal Council (Upper House) on 24 January 2007 (NLB, 2007).

The law, after being approved, legalized the possession of nuclear materials and facilities by other state-owned entities, and provided the creation of a state-owned corporation for all companies involved in the civil nuclear sector, called Atomenergoprom (or Atomprom).

This holding company had several branches, each one responsible for part of the national nuclear industry, except for the nuclear military industry. Atomprom should control the whole nuclear cycle from uranium production through electricity generation, and would oversee nuclear power plant construction in the RF as well as the broad development of nuclear engineering capabilities and scientific institutions (NLB, 2007).

JSC Atomenergoprom

Currently, Atomprom is called *JSC Atomenergoprom* (Joint Stock Company Atomic Energy Power Corporation) that was established to consolidate the assets of the civil part belonging to the Russian nuclear industry. This body absorbed the accumulated experience in the nuclear fuel cycle technology and construction of nuclear power plants which had been practiced in the RF for more than 60 years. The *JSC Atomenergoprom* is controlled by *Rosatom* (Atomenergoprom, 2011).

RosRAO

Rosatom's subsidiary for the management of radioactive waste is RosRAO which began operations in 2009 under a temporary scheme up to completion of radioactive waste management regulation Act (Rosatom, 2010; WNA, 2010a).

According to the Russian "Use of Atomic Energy Act" (dated November 21, 1995) radioactive wastes are nuclear materials and radioactive substances without the possibility of future use (ROSRAO, 2011). Under the law terms, such materials are from operation and decommissioning of the nuclear fuel cycle (mining and processing of radioactive ores, manufacture of fuel elements, nuclear energy production and nuclear fuel reprocessing).

Additionally, according to that Russian legislation, radioactive wastes do not comprise spent nuclear fuel so that not being RosRAO's responsibility but Rosatom's.

Among its activities, RosRAO must provide the radioactive waste management, such as collection and sorting, packaging, storage, transport and burial into geological environment.

Russian President Dmitry Medvedev signed on 11 July 2011 the Federal Law on Management of Radioactive Waste and on Introduction of Changes in Individual Legislative Acts of the Russian Federation, more than one and a half year after the law was introduced in the Russian State *Duma* in December 2009.

The law is a significant first step in establishing a national, central legal framework for radioactive waste management and implements Russia's commitments under the Joint Convention on the Safe Management of Spent Fuel and the Safe Management of Radioactive Wastes, ratified by the Russian Federation in 2006. The law sets out the powers and responsibilities of the Russian Government and federal, regional and local agencies, clarifies ownership of waste as well as storage and burial locations, establishes a national operator for management of radioactive waste, classifies radioactive waste into specific types, establishes the requirements related to management and disposal thereof and places a ban on the construction of new facilities for the disposal of liquid low-level and medium-level radioactive waste in deep geological formations. Implementation of the new law will require adoption of subordinate legislation, which some experts believe may take a few years (NLB, 2011).

Current Situation of Repositories in the RF

No repository of waste is available in the country, although the choice of location should fall in granitic formation of *Kola* Peninsula (in the far north of European Russia, near the border with Finland, part of the *Murmansk Oblast*), where, in the 70's, drillings were performed about 15 km deep into the soil for oil prospection (WNA, 2010a).

In 2003, an area of *Krasnokamensk* town (7000 km east from Moscow, near the Chinese border) was suggested as the site for a major spent fuel repository, based on the fact that the area is the country's largest uranium ore mine (Robinson, 1996; WNA, 2010a).

In 2008, it was presented as a site for a deep geological repository the *Nizhnekansky* granitoid massif in the *Krasnoyarsk* region (capital and the largest city in the Russian territory of the same name, located in western Siberia, in the south-center of the country and one of the USSR's secret sites for the production of war material in the 1950s). Rosatom reported that plans for the construction of facilities for a repository would be delivered by 2015 to the beginning of the project and the creation of an Underground Research Laboratory (URL). The decision on the actual construction is expected in 2025, and, then, the repository will be completed in 2035. In the first phase, the repository should receive about 20 thousand tons of ILW and HLW and must be retrievable (Gupalo et al., 2004; Kudryavtsev et al., 2008; Morov et al., 2009; WNA, 2011).

In the RF, the LLW and ILW are handled similarly to other countries, then, Radon was the organization responsible for the medical and industrial radioactive wastes. It was created in 1961 to develop solutions for those wastes (out of the fuel cycle) and has been responsible for repositories in 16 regions covering the Russian territory. Radon works independently of Rosatom and does not cover wastes from military organizations and nuclear power plants (NM, 2002; Radon, 2010).

In 2010, RosRAO launched plans for a local solution to repositories of power plants and nuclear wastes which are about to be created between 2020-2035 (WNA, 2010a).

The RF has been using injection of liquid waste into deep water-bearing horizons (depth of the wells was up to 1500 meters deep) and injection of liquid waste/cement mixtures into fractures induced in impermeable strata (by hydraulic fracturing) (Pimenov et al., 1980; WNA, 2010a).

In 2008, attempts to experimental plans to build four to six LLW and ILW regional repositories were executed in the northwestern region of the country, Eastern Europe and South regions of Ural Mountains - border between Europe and Asia. For waste containing long-lived radionuclides, the establishment of one or two repositories in Siberia and southern part of the Ural Mountains have been planned too (WNA, 2010a).

The RF had great interest in hosting an international HLW repository and, in 2001, the Russian parliament passed legislation to allow imports of nuclear fuel irradiated. Then, President Vladimir Putin signed an Act creating a special committee to approve and supervise those imports. However, the country did not support or accept external control from international communities because of its ties with countries like Iran. In this understanding, in July 2006, Rosatom announced that it would not continue to accept spent fuel from foreign origin (McCombie and Chapman, 2004; WNA, 2010a).

INDIA

India is not part of the Non-proliferation of Nuclear Weapons Treaty - NPT from United Nations (which entered into force on March 5, 1970) because of its weapons program. So, 34 years ago, the country was excluded from trade reactors for power plants and nuclear materials, which has hindered its civil nuclear energy development by 2009 (WNA, 2010b).

India's position against the NPT, as the country officially has weapons, has limited its ability to participate in international cooperation in the nuclear energy field and caused the junction of its military and civil installations once the country's nuclear resources have been limited. India has always refused to sign the NPT, arguing that the treaty was discriminatory and would not completely ban nuclear weapons in the world. The strategic and geographical location of India, along with their experiences with neighboring countries, has influenced his argument about its complete disarmament.

Because of these trade bans and lack of uranium, India has developed a single nuclear fuel cycle based on thorium.

The Indian nuclear program had its origins in the early 1940s and, since then, has grown to a considerable size in scope and content, with facilities and activities spread throughout the country, affecting the social, economic life and nation's politics. India has effectively used nuclear power for the evolution of society, especially in the agriculture and medicine field. It is a country that insisted on technology and its own resources, emphasizing the importance of self-reliance. This resulted in the development of an industrial necessary backbone for a nuclear energy program (Mannully, 2008).

Regarding the legislative framework, India has advocated the peaceful use of nuclear energy, which has been governed by the Atomic Energy Act of the country. The first version of that law was passed in 1948, but was soon revoked in 1962, in favor of more detailed decrees and comprehensive items "to provide for the development, control and use of atomic energy in favor of the Indian people's welfare and for other peaceful purposes" (Mannully, 2008). Mannully (2008) also points out that the current law allowed the central government to carry out all the tasks associated with the nuclear energy use, thus, the Indian nuclear program has been completely ruled by government agencies. This is because the central government has enjoyed exclusivity and control over all matters relating to the nuclear field.

Kudankulam Nuclear Power Plant (or Koodankulam NPP or KKNPP) is the single largest nuclear power station in India, situated in *Koodankulam* in the *Tirunelveli* district of the southern Indian *Tamil Nadu* state. The construction of the plant began on 31 March 2002. KKNPP is an Indo-Russian joint venture located in India's southern of *Tamil Nadu* state, which is currently finalizing the construction of the first two units, KKNPP units 1 and 2, as both countries have plans for more two units in the near future. KKNPP units 1 and 2 were initiated by the signing of an agreement by the Prime Minister of India and the President of the Union of Soviet Socialist Republics (USSR) on 20 November 1988 (NLB, 2012).

DAE

In August 1948, under the Atomic Energy Act, the Indian Atomic Energy Commission (AEC) was established in Department of Scientific Research founded in June 1948. The Department of Atomic Energy (DAE) was created on August 3, 1954 under the Prime Minister Jawaharlal Nehru's administration. According to a government standard dated March 1, 1958, AEC was established into the DAE (AEC, 2011).

DAE's objectives were to develop technology, research and the operation of commercial reactors. It is important to say that the current Indian atomic energy law is from 1962 and it only allows government offices to be involved in the nuclear industry. The Indian Atomic Energy Commission is the main nuclear political body in the country (WNA, 2010b).

The Atomic Energy Establishment was set up in *Trombay*, a suburb northeast of the *Mumbai* city in Maharashtra state (on the southwest coast of the country, about 300 km from New Delhi capital) in 1957 and renamed as *Bhabha* Atomic Research Centre (BARC) 10 years later (WNA, 2010b; BARC, 2011).

AERB

The Atomic Energy Regulatory Board – AERB was established in 1983 subjected to the Atomic Energy Commission, but it is independent of DAE. AERB was responsible for the regulation and licensing of all nuclear installations and their safety and has the authority conferred by the Atomic Energy Act for protection against radiation. However, AERB has not been an independent statutory authority and, in 1995, a report on DAE's facilities safety assessment was filled by AEC, thus, showing power of decisions coming from AEC. In April 2011, the Indian government announced that it would make laws for the creation of a new Nuclear Regulatory Authority, being completely independent and autonomous, therefore, replacing AERB.

On February 3, 1987 entered into force the Safe Disposal of Radioactive Wastes Rules 1987, which dealt on radioactive wastes and basically provided licensing authorization procedures and definitions of technical terms (DAE, 1987).

Current Situation of Repositories in India

Concerning for the environment and establishing radiation protection goals are among the top priorities in the planning of India's nuclear program. Currently, there are seven surface repository facilities for LLW and ILW, which are operating near nuclear reactors in various parts of the country. These repositories are routinely subjected to monitoring and safety / performance evaluation. An interim storage facility is in operation to store vitrified packs for 30 years or more. India has sought other regions with specific rock characteristics to fulfill its geological repository program, for a preliminary design of a URL (Underground Research Laboratory) repository, and technological research for packaging waste in ceramics and glass matrix in a repository simulated geological have been successful. DAE, with the participation of the Indian industry, has developed remote devices necessary for operation and maintenance of a waste management system (Raj et al., 2006).

In India, the radioactive wastes from nuclear reactors and reprocessing plants are processed and stored at each site of these nuclear facilities. Waste immobilization plants are in operation in *Tarapur* (district in the Indian state of *Maharashtra*) and another is being built in *Kalpakka* (city on the southeast coast of the country in *Tamil Nadu* State). The research on the location and disposition of HLW in a geological repository is in progress by BARC.

PEOPLE'S REPUBLIC OF CHINA (PRC)

When China began to develop nuclear energy, a strategy for the nuclear fuel cycle was also formulated, which was announced at a IAEA's Conference in 1987. The activities towards the spent fuel involved in-reactor storage and storage with reprocessing far from the reactors. China National Nuclear Corporation - CNNC, issued a state regulation on the treatment of civil spent fuel as a basis for a long-term government program. There is a 41 cent fee US$ tax per kilowatt hour focusing on the production of spent fuel for finance management, reprocessing and disposal of HLW (WNA, 2010c).

According to Shiguan (1987), the PRC with its various ministries have compiled nuclear legislation since 1982, giving priority to safety and quality, and a regulatory system divided into two categories. The first concerning to the administrative regulations and the second on regulations, criteria and standards subject to the Chinese Atomic Energy Act that has stated rules on research and development, uranium mining, materials control, nuclear facilities, radiation protection, radioisotopes, transport and compensation for nuclear damage.

NNSA

The National Nuclear Safety Administration (NNSA) or China's Nuclear Regulatory Commission (NRC) or China Nuclear Regulatory Commission (NRC) was established in 1984 under the name of the State Commission of Science and Technology in order to exercise control over civil nuclear facilities, including safety regulation and guarantee for the safe development of peaceful uses of nuclear energy. The NNSA established the Nuclear Safety Centre in Beijing in order to provide technical assistance to regional bodies in areas where nuclear facilities are located. The Nuclear Safety Advisory Committee, which was established in 1986, is another agency that provides the NNSA licensing plans, research and development conditions planning and nuclear safety policy (NLB, 1989a).

On October 29 1986, PRC promulgated regulations on safety supervision and control of nuclear civil installations that centralized the NNSA supervision throughout the Chinese territory (NLB, 1987a).

Shortly, the rules established that from design to decommissioning of any civil nuclear facility, it was necessary to protect the safety of workers and the environment and also promote the minimization of the incident effects. The NNSA, as a responsible body for the licensing of nuclear facilities as well as legislator and inspector of safety at civil nuclear facilities, accumulated responsibility: for creating research and development of institutes and regional offices; for establishing international contacts on nuclear safety; for promoting public information, personnel training; and for devising an advisory committee about nuclear safety. It can also punish with sanctions, revocations of licenses and suspensions (NLB, 1987a; NTI, 2004).

CAEA

The NNSA is subordinated to the China Atomic Energy Authority (CAEA). It has been established in 1984 as licensing and regulatory body and also has to execute international agreements on safety, reporting directly to the State Council. NNSA is sufficiently independent of CAEA, which has planned capacity and viability of nuclear power plants in China.

In December 1995, it was published by the Chinese government a collection of legislation on regulations established by NNSA in the nuclear safety field, which was named Collection of Regulations on Nuclear Safety of the People's Republic of China. For example, within several regulations in the field of nuclear materials control, nuclear emergencies, it was necessary to point out in this compendium, the existence of complementary codes, such as the safety code on Radioactive Waste Management of Nuclear Power Plants dated August 29 1991 and named HAF0800. The Code sets out the safety principles for waste management, covering the responsibilities of the operating organization and relevant authority, the

Evolution of the International Law on Wastes Disposal ... 165

management system of the transport and disposal of radioactive wastes and the management of radioactive wastes generated from decommissioning and nuclear incidents (NLB, 2000).

CNCC

On May 4 1982, the Ministry of China Nuclear Industry was created from a second Ministry of Machine-Building, being reorganized six years later with the name of China National Nuclear Corporation (CNNC) (NTI, 2011a).

The CNNC, created on September 16 1988 by the State Council authority, has its President and Vice President appointed by the State Council Premier. It is an economically self-sustaining enterprise and not a government administrative body (NTI, 2011a).

On July 1 1999 it was approved by the State Council the establishment of two state-owned companies in the Chinese nuclear industry. Then, CNNC remained responsible for the promotion and development of nuclear energy and it was divided into two distinct groups of companies, namely China Group Corporation Nuclear Industry (which remained, however, the acronym CNNC) and China Nuclear Engineering & Construction Corporation (CNEC) (NLB, 2000).

The larger of the two groups, the China Nuclear Industry Group Corporation - CNCC, became a state-owned conglomerate that now controls all nuclear issues outside the construction industry and it was composed of 246 companies and institutions (NLB, 2000).

CNEC became a state-owned enterprise, under the direct supervision of the State Council, then, CNEC was replaced by CNNC over the nuclear construction sector, which was composed of 13 separate business units (NLB, 2000).

The former CNNC's administrative functions have been transferred to the State Commission of Science, Technology and Industry of National Defense. The China Atomic Energy Authority, which is an integral part of this Commission, has been responsible for the management of the peaceful uses of nuclear energy and the promotion of international co-operation (NLB, 2000).

Currently, CNNC oversees all civil and military aspects of nuclear PRC's programs, "combining military production to civil production, and the nuclear industry as a basis for the development of nuclear energy to promote a diversified economy" (NTI, 2011a).

EEEC

In PRC, the CAEA is the government organization responsible for plans development and projects for the HLW disposal. The Ministry of Environmental Protection, its affiliated institutes and NNSA are regulators. The implementation of activities related to the radioactive wastes disposal is currently managed by CNNC.

CNNC is the operational authority, responsible for site selection, construction and operation of repositories for each region. The Everclean Environmental Engineering Corporation (EEEC), a CNNC's subsidiary, was founded to deal with LLW and ILW solids (NTI, 2011b).

In 2006, PRC's government along with CAEA, the Ministry of Environmental Protection and the Ministry of Science and Technology, with the aim of building a HLW repository in 2050, published the program in "2006 R&D Guidelines for Geological Disposal." This document consisted of three steps for the HLW disposal, namely:

(1) between 2006-2020: studies at URL (Underground Research Laboratory) and choice of location for a HLW repository;
(2) between 2021 to 2040: underground "in loco" testing; and
(3) from 2041 to 2050: the construction of the repository and its operation (Wang, 2009).

Current Situation of Repositories in PRC

China's current policy of spent fuel management is the interim storage inside or outside the nuclear reactor facilities.

In October 1994, the IAEA sponsored the "Conference on practices and issues to develop radioactive waste management in PRC" and CNNC announced a policy about the burial location for LLW and ILW solids in deep ground layers, and permanent burial of HLW. Due to the nature of that policy, all provinces and municipalities in PRC should establish interim waste storage, since all nuclear facilities have their own storage and treatment units, with 21 such facilities in PRC (NTI, 2011b).

The site selection for a HLW repository has had four stages: national, regional, district and local. Screening for the location of the site began in 1986 with a focus on *Beishan* area in northwest China in *Gansu* Province. Other most promising districts were *Quinhongquan* and *Jiujin* in the southern part of the region (NTI, 2011b).

A centralized spent fuel storage was built in *Lanzhou* Nuclear Fuel Complex, 25 km northeast of the city and capital of the same name in *Gansu* Province. The initial phase of this project had a storage capacity of 550 tons of waste but could be doubled in the most widely used fuel stored in the reactor (WNA, 2010c; NTI, 2011b).

HLW will be vitrified, encapsulated and placed in a geological repository about 500 meters deep. The site selection and evaluation is ongoing since 1986 and is focused on three candidate sites in *Beishan* area and should be completed by 2020. The locations have granite formation and URL will be built between 2015-2020 to operate for 20 years; all procedures are to be taking place under the "2006 R&D Guidelines for Geological Disposal" (Wang, 2009; WNA, 2010c).

Six deep wells and six shallow wells were drilled at three locations in *Beishan* site during the period 2000-2009. The results from studies have showed that the granite rock formation have high integrity, low fracture density and low hydraulic conductivity, indicating that the site is potentially useful for the construction of future geological repositories (Wang, 2009).

168   Daniel Figueira de Barros and Daniel Marcos Bonotto

Despite the progress made in many ways, the Chinese task towards to HLW has found many engineering, social, economic, scientific and technical challenges to be faced (Wang, 2009).

SOUTH AFRICA

Like Brazil, South Africa has two nuclear reactors - *Koeberg 1* and *2*, located 30 km north of Cape Town, near the town of *Melkbosstrand* on the west coast. They account for about 5% of the electricity production in the country (WNA, 2010d).

AEC

In 1948, the Atomic Energy Act of South Africa created the Atomic Energy Council, which later became the Atomic Energy Corporation of South Africa Ltd. (AEC) (WNA, 2010d).

In 1963, the Nuclear Installations Act demanded licensing of nuclear facilities and, in 1982, under the Atomic Energy Act, the AEC was responsible for all nuclear issues, also including uranium enrichment. An amendment to this law created the Council for Nuclear Safety (CNS), an autonomous entity responsible for the licensing of nuclear facilities (WNA, 2010d).

CNS

On June 1 1988, the Atomic Energy Act was amended in order to set CNS as a legal entity, thus, being more than a mere advisory body for the AEC and Minister for Mineral and Energy Affairs (MMEA). Consequently, it may give its opinion on licensing, health and safety issues in nuclear facilities during the production, use, storage and disposal of nuclear materials. AEC was a state-owned public company, responsible for nuclear

Evolution of the International Law on Wastes Disposal ... 169

development and energy production that did not require licensing for its nuclear activities. CNS should start issuing licenses from this new 1988 law (NLB, 1989a).

On 26 September 1993, a new amendment entered into force on the Atomic Energy Act, this time revoking and replacing the 1982 Act and changing the Hazardous Substances Act No. 15 of 1973 that only redefined as nuclear materials those consisting of or containing natural or artificial radionuclides (NLB, 1979, 1994).

The Act also brought references to other issues such as liability, indemnity and patents among other topics. It is important to note that such Act dealt with the control over the radioactive wastes disposal and spent fuel storage, leaving this task in charge of AEC, as well as nuclear technology development functions, trade and implementation of international safeguards agreements (NLB, 1994).

From that Act, AEC had broader power of decisions establishing other companies and international agreements under the MMEA's scrutiny. The organization was headed by a board of directors with a chairman and six other deputies, being financially subsidized by the Ministry of Finance (NLB, 1994).

CNS was responsible for regulating functions and issuing licenses for the construction and operation of nuclear facilities, radioactive wastes disposal, spent fuel storage, among others aimed to authorizations (NLB, 1994).

NNR

The Nuclear Energy Act 1999 granted responsibility to MMEA for nuclear power generation, radioactive wastes management and the country's international commitments (WNA, 2010d).

The National Nuclear Regulator Act of 1999 created the National Nuclear Regulator (NNR), replacing the CNS. In this sense, NNR had responsibility to cover projects of the complete fuel cycle from mining to wastes disposal (NNR, 2010; WNA, 2010d).

It is noteworthy that the National Nuclear Regulatory Act of 1999, when established NNR, aimed to standardize safety standards and regulatory practices for the protection of people, property and the environment against nuclear damage. Thus, this new legislation aimed to draw a clear line between nuclear regulators and the development and use of nuclear materials and equipments, thus, putting each organization responsible for these activities under separate laws (NLB, 2000).

NECSA

In order to create a substitute for AEC, the state-owned South African Nuclear Energy Corporation Limited (NECSA) was created to implement the safeguards agreement, regulate the acquisition, possession, import and export of nuclear fuel, equipment and related materials and even to prescribe measures for the radioactive wastes disposal and spent fuel storage (NLB, 2000; WNA 2010d). NECSA also is liable to deal with the decommissioning and decontamination of nuclear facilities as well as the management of nuclear wastes at national level (NECSA, 2011).

NECSA obtained its mandate from three regulatory frameworks: the Nuclear Energy Act 1999, the 2008 Nuclear Energy Policy and regulations conferred by the Ministry of Energy (NECSA, 2011).

In South Africa, the Department of Energy has overall responsibility for nuclear energy and regulates its rules under the Nuclear Energy Act 1999, National Radioactive Waste Disposal Institute of 2008 and National Nuclear Regulator (NNR) 1999 (ENERGY, 2012).

The Department of Environmental Affairs is responsible for the environmental assessment of projects, and a cooperation agreement with the NNR for nuclear projects (WNA 2010d; ENERGY, 2012).

On January 9 2009, the Act No. 53 of 2008 published the National Radioactive Waste Disposal Institute Act, with the aim of creating a state-owned institution in order to: design and idealize waste disposal solutions; manage, operate and monitor waste disposal facilities; design and build repositories; help generators of small waste amounts; and maintain a

Evolution of the International Law on Wastes Disposal …   171

database of the country amounts of wastes (SAGO, 2009). The institute will be primarily financed by fees paid by waste generators and funds coming from the Parliament and should maintain a radioactive waste management fund (SAGO, 2009).

Current Situation of the Repositories in South Africa

A program to select a suitable site for the nuclear wastes disposal involved the examination of a variety of socio-economic parameters related to the geology of large areas in South Africa in 1978 (NLM, 2005).

The initial phase of the investigation culminated in 1983, when three farms that are now the *Vaalputs* repository, were acquired by the State on behalf of NECSA, which has been responsible for its management (NLM, 2005).

The National Institute of disposal of radioactive waste has not been realized yet, and the responsibility for the disposal of nuclear waste has been under NECSA (NLM, 2005; WNA 2010d).

NECSA is operating the national LLW and ILW repository in *Vaalputs* in *Springbok* town, Northern Cape province at 923 km southwest of Pretoria. The repository was commissioned in 1986 to receive wastes from nuclear power plant *Koeberg 1* and *2,* and it is funded by fees paid by state-owned *Eskom* Electricity. Some LLW and ILW from hospitals, industry and NECSA are arranged in a site in *Pelindaba* (at 33 km west of Pretoria), which is a nuclear research center, where South Africa has developed and built nuclear bombs, which are now stored in this site (WNA, 2010d).

It should be noted that spent fuel was buried in trenches 10 meters deep, which were filled with compacted soil and rehabilitated with the local flora planting. In June 1997, it was revealed that some discovered barrels containing wastes were rusty and were leaking radioactivity. NNR temporarily suspended operations until the licensing conditions were met by the operators and reported it to the local communities in September 2003 (Fig, 2005).

CONCLUSION

The core of a legal system is its protective function and its continuity and this should be properly applied in the case of radioactive wastes which require long-term planning and continuity in its regulatory. Establishment of specialized agencies, constitution of funds, legal provisions on liability and compensation for nuclear damage that the nuclear industry is obliged to comply with is essential for the nuclear field.

All the countries under this study have published several legislation about wastes and geological repositories, as well as creating, altering and extinguishing several agencies in the course of the evolution of legislation and government policies. BRICS countries have followed a similar trend with the regulation of geological repositories concentrated in the Federal Government – central government - yet having a single agency with full control of actions, attributions, responsibilities and tasks to legislate in the field of wastes and geological repositories.

It is believed that with the decentralization of the nuclear monopoly in favor of several public or private and/or privatized entities, financial and supervision factors might be diluted. In this sense, there might be a huge saving of public researches, hiring charges for the public sector and immediate advances in the nuclear waste management sector and its disposal in repositories.

Considering the mutability of law itself and the regulators of the nuclear sector, it is difficult to establish legislation and ensure its compliance by the next generations, so the subject of legislation on geological repositories needs more debate, research, consistency and study. Then, responsibility of each nation should not be limited to today, but also to tomorrow, which is extended to an uncertain future, because wastes stored today, will influence the environment and generations to come.

ACKNOWLEDGMENTS

D.F.B. is deeply thankful to CAPES (*Coordenação de Aperfeiçoamento de Pessoal de Nível Superior*), Brasília/DF, for the scholarship that supported his Master Dissertation and PhD Thesis, both conducted at the Graduated Program in Regional Geology, IGCE-UNESP, Rio Claro, São Paulo, Brazil.

REFERENCES

AEC (Atomic Energy Commission) (2011). Government of India. [http://www.aec.gov.in].

Atomenergoprom JSC (2011). Joint Stock Company Atomic Energy Power Corporation Atomenergoprom. [http://www.atomenergoprom.ru/en].

Balzani, V., & Armaroli, N. (2011) *Energy for a Sustainable World: From the Oil Age to a Sun-Powered Future* (2nd ed.). Weinheim, Germany: Wiley-VCH.

BARC (Bhabha Atomic Research Centre) (2011). [http://www.barc.ernet.in/about].

Barros, D. F. de (2012). *Estudo comparativo da evolução da legislação internacional e brasileira sobre repositórios geológicos de rejeitos radioativos.* Tese de Doutorado. Rio Claro, SP: UNESP. [Comparative study of the evolution of the international and Brazilian legislation concerning to geological repositories of radioactive wastes. PhD Thesis. Rio Claro, SP: UNESP].

Barros, D. F. de, & Bonotto, D. M. (2015a). Major features on Brazil's legislative policy in uranium production and related environmental aspects. In J. A. Daniels (Ed.), *Advances in Environmental Research: Vol. 39* (85-96). New York: Nova Science.

Barros, D. F. de, & Bonotto, D. M. (2015b). Brief overview of the Brazilian legislation and the radioactive wastes geologic repository project. In B. Veress, & J. Szigethy (Eds.), *Horizons in Earth Science Research: Vol. 13* (161-173). New York: Nova Science.

BBVA (Banco Bilbao Vizcaya Argentaria) (2012). [https://www.bbvaresearch.com/KETD/fbin/mult/120215_BBVAEAG LES_Annual_Report_tcm348-288784. pdf?ts=1642012].

DAE (Department of Atomic Energy) (1987). [http://www.dae.gov. in /rules/waste.pdf].

EME (EMerging Equity) (2016). [https://emergingequity.org/2016/01/07/ world-bank-issues-2016-perfect-storm-warning-amid-brics-synchronised-slowdown].

ENERGY (Department of Energy - Republic of South Africa) (2012). [http://www.energy. gov.za/files/nuclear_frame.html].

Fig, D. (2005). Uranium road: questioning South Africa's nuclear direction. [http://books.google.com/books?id=JjNWRmiJ1FUC&pg=PA62&hl= pt-BR&source=gbs_selected_pages&cad=3#v=onepage&q&f=false].

GS (Goldman Sachs) (2010). [http://www2.goldmansachs. com/careers/our-firm/locations/ united-states/history-growth.html].

Gupalo, T. A., Kudinov, K. G., Jardine, L. J., & Williams, J. (2004) Creation and plan of an underground geologic radioactive waste isolation facility at the Nizhnekansky rock massif in Russia. [https://e-reports-erxt.llnl.gov/pdf/313754.pdf].

IAEA (International Atomic Energy Agency) (1980). Regulatory Aspects of Underground Disposal of Radioactive Waste [http://www-pub.iaea.org/MTCD/ publications/PDF/ te_230_web.pdf].

IAEA (International Atomic Energy Agency) (1997). The Joint Convention on the Safety of Spent Fuel Management and on the Safety of Radioactive Waste Management. [https://www.iaea.org/sites/ default/files/infcirc546.pdf].

IMF (International Monetary Fund (2013). [http://www.imf.org/ external/pubs/ft/weo/2013/01/weodata/weorept.aspx?pr.x=91&pr.y=5 &sy=2011&ey=2018&scsm=1&ssd=1&sort=country&ds=.&br=1&c=

223%2C924%2C922%2C199%2C534&s=NGDPD%2CNGDPDPC%2CPPPGDP%2CPPPPC&grp=0&a=].

Kudryavtsev, E. G., Gusakov, Stanyukovich, I. V., Kamne, E. N., Lobanov, N. F., & Beygu, V. P. (2008). Construction of a deep geological disposal facility for final isolation of high-level waste in the Nizhnekansky rock massif. [http://www.iaea.org/OurWork/ ST/NE/ NEFW/CEG/documents/ws022009/4-5.%20Programs%20for%20Deep %20Geological%20Repositories%20and%20Underground%20Labs/4. 7%20Creation%20of%20DGR%20in%20Krasnoyarsk%20Region%20 Engl.pdf].

Mannully, Y. T. (2008). India Nuclear Cooperation and Non-Proliferation. [http://www.oecd-nea.org/law/nlbfr/documents /009_ 026_ArticleMannullyYash.pdf].

McCombie E, C., & Chapman, N. (2004). IAEA Developing and Implementing Multinational Repositories: infrastructural framework and scenarios of co-operation. [http://world-nuclear.org/info/ inf21.html].

Morov, V. N., Kolensnikov, I. Yu, & Tatarinova, T. A. (2009). The Prediction for stressed-deformed conditions of the Nizhnekansky rock massif as a possible zone for disposal of radioactive wastes. [http://www.kscnet.ru/ kraesc/2009/2009_14/art10.pdf].

MRE (Ministério das Relações Exteriores – Itamaraty) (2011). BRICS - Agrupamento Brasil-Rússia-Índia-China-África do Sul. [http://www. itamaraty.gov.br/temas/mecanismos-inter-regionais/agrupamento-brics].

NECSA (South African Nuclear Energy Corporation) (2011). [http://www.necsa.co.za/Portals/1/Documents/4d04001c-20fd-40a5-a871-04daa8447ab5.pdf].

NLB (Nuclear Law Bulletin) (1968). [http://www.nea.fr/law/nlb/NLB-01-EN.pdf].

NLB (Nuclear Law Bulletin) (1979). [http://www.nea.fr/law/nlb/NLB-24-EN.pdf].

NLB (Nuclear Law Bulletin) (1987a). [http://www.nea.fr/law/nlb/NLB-39-EN.pdf].

NLB (Nuclear Law Bulletin) (1987b). [http://www.nea.fr/law/nlb/NLB-40-EN.pdf].

NLB (Nuclear Law Bulletin) (1989a). [http://www.nea.fr/law/nlb/NLB-43-EN.pdf].

NLB (Nuclear Law Bulletin) (1989b). [http://www.nea.fr/law/nlb/NLB-46-EN.pdf].

NLB (Nuclear Law Bulletin) (1991). [http://www.nea.fr/law/nlb/NLB-47-EN.pdf].

NLB (Nuclear Law Bulletin) (1992). [http://www.nea.fr/law/nlb/NLB-49-EN.pdf].

NLB (Nuclear Law Bulletin) (1994). [http://www.nea.fr/law/nlb/nlb-53.html].

NLB (Nuclear Law Bulletin) (2000). [http://www.oecd-nea.org/law/nlb/nlb65.pdf].

NLB (Nuclear Law Bulletin) (2007). [http://www.oecd-nea.org/law/nlb/nlb-79/059-076-National%20legislative%20and%20regulatory%20activities.pdf].

NLB (Nuclear Law Bulletin) (2011). [http://www.oecd-nea.org/law/nlb/nlb88.pdf#page=147 2011].

NLB (Nuclear Law Bulletin) (2012). [http://www.oecd-nea.org/law/nlb/nlb90.pdf].

NLM (Nuclear Liabilities Management) (2005). [http://www.radwaste.co.za/vaalputs.htm].

NM (Nuclear Market) (2002). [http://www.nuclearmarket.com/suppliers/details2.cfm? IDcompany=2386].

NNR (National Nuclear Regulator) (2010). [http://www.nnr.co.za/].

NTI (Nuclear Threat Initiative) (2004). National Nuclear Safety Administration – NNSA. [http://www.nti.org/db/china/nnsa.htm].

NTI (Nuclear Threat Initiative) (2011a). China National Nuclear Corporation – CNNC. [http://www.nti.org/db/china/cnnc.htm].

NTI (Nuclear Threat Initiative) (2011b). China National Nuclear Corporation – CNNC. [http://www.nti.org/db/china/sptfuel.htm].

Pimenov, M. K., Murano, T., Asano, T., & Matsubara, N. (1980). Basic regulatory requirements for carrying out investigations, reasoning and the approving of the disposal of radioactive and other industrial waste in geological formations in the USSR. [http://www-pub.iaea.org/MTCD/publications/PDF/te_230_web.pdf].

Radon (2010). [http://www.radon.ru/].

Raj, K., Prasad, K. K., & Bansal, N. K. (2006). Radioactive waste management practices in India. [http://www.sciencedirect.com/science/articrle/pii/S0029549306000859].

Robinson, P. (1996). Impacts of uranium mining in Krasnokamensk. Environmental damage and policy issues in the uranium and gold mining districts of Chita Oblast in the Russian Far East: A Report on existing problems at Baley and Krasnokamensk and policy needs in the region. [http://www.sric.org/mining/docs/Chitafin.html].

Rosatom (2010). [http://www.rosatom.ru].

RosRao (2011). [http://www.rosrao.ru/wprs/wcm/connect/rosrao/rosraosite/conversion/ nuclear_scrap/].

SAGO (South Africa Government Online) (2009). [http://www.info.gov.za/view/Download FileAction?id=94446].

Shiguan, Z. (1987). Review of Nuclear Legislation. [http://www.nea.fr/law/nlb/NLB-40-EN.pdf].

Tennembaum, J. (2009). *Energia Nuclear: Dínamo da reconstrução econômica mundial*. Rio de Janeiro, RJ: Capax Dei. [*Nuclear Energy: dynamo of the world economic reconstruction*. Rio de Janeiro, RJ: Capax Dei].

Wang, J. (2009). High-level radioactive waste disposal in China. [http://202.127.156.15/ qikan/manage/wenzhang/2010-01-01.pdf].

WNA (World Nuclear Association) (2010a). [http://www.world-nuclear.org/info/default.aspx ?id=26576&terms=russia].

WNA (World Nuclear Association) (2010b). [http://www.world-nuclear.org/info/default. aspx?id=338&terms=india].

WNA (World Nuclear Association) (2010c). [http://www.world-nuclear.org/info/default. aspx? id=26187&terms=china].

WNA (World Nuclear Association) (2010d). [http://www.world-nuclear.org/info/default.aspx?id=372&terms=south%20africa#Note_j].

WNA (World Nuclear Association) (2011). International Nuclear Waste Disposal Concepts. [http://world-nuclear.org/info/inf21.html].

Xinhuanet (2016). [http://news.xinhuanet.com/english/china/2016-10/18/c_135762265.html].

In: Radioactive Wastes and Exposure
Editor: Austin D. Russell

ISBN: 978-1-53612-213-8
© 2017 Nova Science Publishers, Inc.

Chapter 5

ASSESSMENTS OF LOCAL ACCEPTANCE OF RADIOACTIVE WASTE FACILITIES

Taehyun Kim, PhD[*]
Korea Environment Institute, Sejong, Republic of Korea

ABSTRACT

The purpose of this chapter is to introduce studies on the assessment of local acceptance of radioactive waste facilities (RWFs) using different methods. Because radioactive waste disposal facilities (RWDFs) are one of the most controversial locally unwanted land uses, siting these facilities near human habitation has been a growing issue in urban planning and environmental management. Some studies on the siting issue have been conducted using engineering and geographical approaches. Others considered social and economic perspectives using quantitative and qualitative methods. However, there is a lack of literature combining these different approaches. Four studies introduced in this chapter measured local acceptance for low- and high-level RWDFs using qualitative, quantitative, and mixed methods approaches.

The first two studies analyzed the spatial patterns of the referendum results for siting a low-level RWDF. The facility was assigned to be placed

---

[*] Corresponding author: kimth@kei.re.kr

in Gyeongju city after a competitive local referendum amongst four candidate cities in Korea in 2005. However, many conflicts between the residents living within and near Gyeongju occurred after the decision. By analyzing spatial patterns of the referendum data, the first study identified that the local acceptance near and far from the facility were clustered with different values. The results of face-to-face interviews showed that people near the nuclear power plant had low risk perception, and the benefit of financial compensation for the districts offset the cost of potential risk. The second study analyzed the spatial distribution of the acceptance rate in each ward of the four candidate cities using several spatial statistical methods and several types of interviews. The results showed that the referendum system had a problem with spatial inequity within and across its jurisdiction.

The last two studies examined the local acceptance for a high-level RWDF using qualitative and quantitative methods. As the temporary storage for spent nuclear fuel in Korea has almost reached its limit, the question of relocating the facility has become urgent. Because perceptions on high-level radioactive waste may be higher compared to those on low-level radioactive waste, the latter two studies identified perception types and factors of local acceptance for spent nuclear fuel repository. One study identified four types of local acceptance—safety concerns–government distrust, safety trust–government trust, safety concerns–conflict avoidance, and citizen participation—and differences among these perception types using Q methodology. The other study conducted a survey and identified five factors of residents' perception of spent nuclear fuel repository by analyzing the structural equation model. The results showed that environmental impacts and economic feasibility had a high positive relation to local acceptance rather than risk perception. In particular, environmental impacts were distinctively high regardless of demographic characteristics.

These examples may provide new perspectives on management strategies of locating RWFs regarding local acceptance of these facilities.

Keywords: local acceptance, radioactive waste facilities, spatial analysis, Q methodology, Korea

1. INTRODUCTION

The purpose of this chapter is to introduce studies on the assessment of local acceptance of radioactive waste facilities (RWFs) using different methods. Because radioactive waste disposal facilities (RWDFs) are one of the most controversial locally unwanted land uses (LULUs), siting these facilities near human habitation has been a growing issue in urban planning and environmental management. Some studies on the siting issue have been conducted using engineering and geographical approaches. Others considered social and economic perspectives using quantitative and qualitative methods. However, there is a lack of literature combining these different approaches. Four studies introduced in this chapter measured local acceptance for low-/high-level RWDFs using qualitative, quantitative, and mixed methods approaches.

2. SPATIAL ISSUES OF SITING RWDFs

2.1. Study 1: "Analysis of Spatial Patterns of the Result of Referendum for Siting a Radioactive Waste Disposal Facility in Gyeongju, Korea"

Low-level RWDFs are one of the most challenging LULUs. One proposed low-level RWDF in Korea has failed to find a site for almost 30 years (since 1984) due to public resistance. In 2004, to facilitate siting, the Korean government promised three major incentives for the city chosen: 1) it would provide financial support, 2) it would locate the headquarters of the Korea Hydro & Nuclear Power Company in the city, which could revitalize the local economy, and 3) it would not locate a high-level RWDF in the city. In 2005, the decision was made to locate the facility in Gyeongju because the city had the highest approval rating amongst the local referendums in the four candidate cities. The decision process used to site the low-level RWDF appeared to have been successful. However, siting LULUs via such a competitive local referendum, whose main prerequisites are economic rewards for a whole city, involved geographical problems related to spatial justice.

In this context, this study suggested the following two research questions, focusing on the spatial difference in the results of the competitive referendum, which presumed regional rewards to the regional government: 1) If approval or disapproval is a result of personal judgment, the approval rate would not exhibit any spatial pattern. If the neighborhood effect, which influences the judgment of individuals at a regional level, does exist, the approval rate of certain regions will show spatial patterns up to a certain degree. Additionally, a region may display outliers, distinguishing it from adjacent regions. Do such patterns really exist? If so, how do they look? 2) When the approval rate is concentrated in a certain region or a region shows outliers, is it because there was any external shock factor? Was there any special socio-cultural reason, unlike other regions? Are there any unique geographical or topographical characteristics?

The objective of this study was to investigate spatial patterns and to identify their causes by analyzing the spatial distribution and exploring the social and economic backgrounds of regions exhibiting outliers. To this end, a mixed methodology was applied, which employed both the quantitative approach effective for objective hypothesis testing and the qualitative approach suitable for exploring possible causes of the problem.

This study selected the city of Gyeongju, which hosted the RWDF with the highest approval rate, as the study site among the four local governments, Gyeongju (GJ), Pohang (PH), Yeongdeok (YD), and Gunsan (GS), that had voted in a referendum to locate an RWDF (Figure 1).

In 2005, the population of Gyeongju was 276,060 and the voter population 209,099. In the results of the referendum in the city, the recorded voting rate was 70.8% and the approval rate 89.5% from 76 voting precincts located across 25 *eup*, *myeon*, and *dong* within Gyeongju. The site for RWDF construction, Bonggil-li in Yangbuk-myeon, is geographically located in a mountainous area along the coast, east of the city and is adjacent to the Wolsong Nuclear Power Plant. The site visit revealed that the neighboring residential areas have low visibility to the construction site as it is topographically located in the mountain. Thus, the RWDF is generally not visible from those residential areas unless people visit the site by car.

Assessments of Local Acceptance of Radioactive Waste Facilities 183

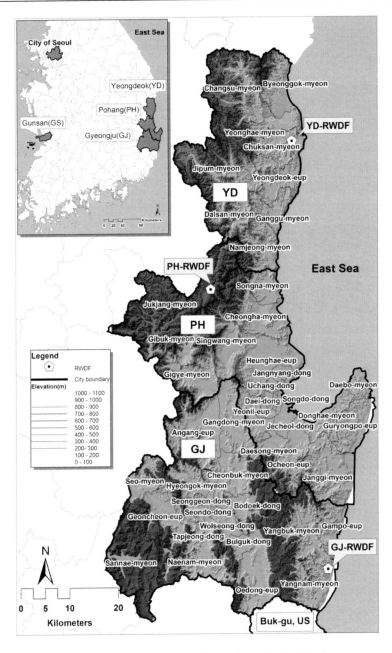

Figure 1. Topographical map of candidate cities (Kim and Kim, 2014).

The referendum results were used to measure residents' local acceptance of the RWDF. This was because it is difficult to draw a suitable sample representation of each region when measuring local acceptance by a survey. Furthermore, it also held the risk of distorting the real position of respondents during measurement, as the study aimed to analyze the characteristics of the geographical distribution of local acceptance.

The election commission of each candidate city provided the referendum results, including the number of votes, the votes for and against the siting, invalid votes, abstention for each of the voting precincts, and the absentee ballots. The election commission also provided the name and address of each of the voting precincts. The distance (in meters) from the location of the proposed site to each voting precinct was calculated using an electronic map. The referendum geography used in this study partitioned the city (*si*) or county (*gun*) into several towns (*eup*), townships (*myeon*), blocks (*dong*), and districts (*li*). These political subdivisions (towns and blocks) were further subdivided into several electoral zones (voting precincts) by the National Election Commission; the electoral zones were based in part on population and proximity.

In this study, the spatial scale was divided into *eup*, *myeon*, *dong*, and voting precincts. Moran's I was used to test the neighborhood effect in each spatial scale. Getis-Ord Gi* was used for each voting precinct to identify regions showing outliers forming patterns with relatively higher or lower approval rate values. Getis-Ord Gi* requires distance input, which defines neighbors. In this study, the distance was set as Minimax, where spatial data had at least one neighbor.

As for voting precincts, which showed a significant difference in their approval rate compared to that of other regions based on the spatial statistics analysis results, a survey was conducted in May and December 2008. Persons who were considered to be well aware of the situation of each region and able to represent the local opinion—*li* (district) heads, *tong* (sub-district within *dong*) heads, representatives of independent environmental watches, civil servants working at *eup*, *myeon*, and *dong* offices, professors, *si* (city) council members, and *gun* (county) council members—were surveyed with open questions via telephone, visits, and group interviews.

To investigate the causes of the outliers found in the spatial analysis, the following question was posed: "Was there any special reason for the particularly high (or low) approval rate in the region you belong to?" Responses were collected through voice-recording or by taking text notes. The contents of media interviews were used as secondary data concerning the local referendum on the RWDF. In the group interviews, the representatives of each town shared their opinions concerning the position and situation of the local residents at the time of the 2005 local referendum and any changes brought to the regions neighboring the selected site as of 2008, three years since the decision was made on siting the facilities.

The local referendum results were spatially analyzed. It was found that the approval rates of the regions (*eup*, *myeong*, and *dong*) far away from the RWDF were generally higher. However, those of voting precincts nearest to the selected candidate site for the RWDF were higher in the analysis by voting precinct. The biggest problem in deciding on the site for hazard facilities via local referendum was spatial inequity in setting the voting precincts and the economic rewards. The decision made on siting the facility caused local conflicts among residents of Gyeongju and subsequent conflicts with the city of Ulsan. In regions such as Gyeongju, which are topographically divided, the local acceptance may vary from area to area, and internal, local-level conflicts may arise as a result.

As was well demonstrated by the case of Gyeongju, conflicts may arise between the urban and rural areas concerning rewards when the location of hazard facility is assigned in a rural town or fishing village housing relatively fewer people within the same administrative region. Farmers and fishermen engaged in agriculture, livestock industry, and fishery are dependent on personal connections for maintaining their livelihoods. These groups are often characterized by old age and also make up the social class marginalized from the benefits of social infrastructure and cultural facilities. Agreeing to locate undesirable facilities based on rewards is not really a matter of choice for them. It is an inevitable dilemma in which they have to accept it for a better life even though the acceptance may eventually be unfavorable for them. Thus, the case of Gyeongju implies that the difference in the level of intra-regional development may actually further fuel the conflicts concerning the problem of siting undesirable facilities.

The existing studies therefore provide a useful basis for making a decision on siting hazard facilities using spatial geographic information. However, their limitation is that the conflict factors innate in society were not taken into consideration. The analysis approach introduced in this study and the application of spatial geographical information and interview surveys not only helped decision making considering physical safety but also predicted potential conflict factors originating from topographical and regional characteristics. Finally, this approach also provides crucial information in decision making that reflects social consensus.

2.2. Study 2: "The Spatial Politics of Siting a Radioactive Waste Facility in Korea: A Mixed Methods Approach"

This study sought to identify whether there were any spatial inequity problems in the procedure used to site LULUs at various locations. The purpose of this study was to explore the spatial matters associated with the referendum system used for the siting of an RWDF and to determine the reason why the conflicts continued after a decision had been made. Analyzing this example of a competitive local referendum for siting a low-level RWDF in Korea could aid future decision-making processes used in the siting of LULUs. This study investigated how these assumptions were applicable at the time of voting in the four candidate cities.

Papers on the regional acceptance of RWDFs have discussed the issues of spatial equity and financial compensation (for example, Choi, 2008; Chung & Kim, 2009; Chung, Kim, & Rho, 2008). However, these studies involved quantitative research based on surveys and individual analyses. Despite the development of decision-making software that includes geographical information and spatial analysis techniques, few studies have focused on the relationship between spatial factors and regional acceptance. Moreover, there have been studies that have used engineering and geographical approaches (for example, Kontos, Komilis, & Halvadakis,

2005; Stoffle et al., 1991; Stone, 2001), and others that considered social and economic perspectives on the siting issue (for example, Caplan, Grijalva, & Jackson-Smith, 2007; Choi, 2008; Chung et al., 2008; Kunreuther, Kleindorfer, Knez, & Yaksick, 1987; Rogers, 1998; Sjoberg, 2004) using quantitative (for example, Maderthaner, Guttmann, Swaton, & Otway, 1978; Swallow, Opaluch, & Weaver, 1992; Warren et al., 2005) and qualitative methods (for example, Lidskog, 2005; Van der Horst, 2007), but there is a lack of literature combining these four approaches. This research aimed to verify spatial clustering patterns and to explore the social and economic backgrounds of regions that show particular characteristics. A mixed methodology was employed in this study. The methodology included both a quantitative approach that is effective in verifying the hypotheses objectively and a qualitative approach to properly explore the causes of the problems.

Spatial statistics, including the Getis-Ord General G and Gi*(Getis & Ord, 1992), were employed to identify spatial patterns of similarity and outliers in votes, respectively. For the outliers, individual and group interviews were conducted face-to-face or by telephone to understand why specific regions displayed differences compared to the neighborhood consensus with regard to acceptance; the interviews were performed from May 3 to December 4, 2008.

The interviews focused on local officials of the towns that showed an unusual spatial pattern in the acceptance rate, because such officials might represent the general opinions of the residents in each town, and they might be able to explain why the residents supported or did not support the nuclear facility at that time. The local names of subdivisions for three adjacent candidate cities (YD, PH, and GJ) are shown on a topographical map in Figure 1.

The level of acceptance varies with experience, topography, level of development, and boundaries that cause spatial effects. Compensation and a decision-making process that disregard these factors may induce conflicts in inter- and intra-regional relationships. Because the unwanted facility must be located in a rural area with a comparatively low population within its administrative district, conflicts between rural areas and the city may arise over compensation. People who rely on agriculture, livestock, and fisheries to earn their living and who depend on the neighborhood environment are mostly elderly and marginalized with regard to the benefits of social infrastructure and cultural facilities. The compensation associated with siting the unwanted facility represents a dilemma for them; they have no other alternatives but to accept the facility, even though it could be disadvantageous to the region and their livelihoods. Thus, the difference in the level of intra-regional development may amplify conflicts over the siting of an opposed facility.

If geographical and regional characteristics affect voting behavior related to the siting of an RWDF, the model suggested by this study can be employed to anticipate future regional conflicts. As Gallagher, Ferreira, and Convery (2008) discussed, however, without sufficient consultation and comprehension regarding an unwanted facility before and after the compensation is presented, complaints about unequal compensation and anxiety over the potential risks that may arise during the transportation of radioactive waste can reduce acceptance and cause conflict within and between regions.

As Korea is in the process of publicizing the issue of siting a high-level RWDF, topographically safe and socially acceptable locations should be selected. Spatial analysis of the previous referendum would help to determine whether the issue of spatial equity in a competitive referendum that is conditional upon regional compensation for the entire county was problematic.

3. LOCAL ACCEPTANCE OF RWF

3.1. Study 3: "A Study on the Perception Types and Characteristics on Spent Nuclear Fuel Repository: Focused on Q Methodology"

As an RWF, spent nuclear fuel repository is considered a LULU due to concerns about risk, health damage, and facility safety. Spent nuclear fuel repository has almost reached its maximum capacity in Korea and location decisions for the facility has become an urgent issue. Even though severe conflicts are expected in the process of choosing a location, little research has been done to investigate opinions of individuals dwelling around the affected areas. Therefore, this study aimed to investigate the residents' perception types and characteristics of spent nuclear fuel repository using Q methodology. It identified four types of spent nuclear fuel repository by conducting a Q survey with 45 statements to 56 people.

In detail, the following three study problems were suggested: First, how is the perception on the attitude, belief, confidence, and value of a variety of surveyed respondents concerning the spent nuclear fuel repository distinguished? Second, what are the characteristics of each type of perception of spent nuclear fuel repository? Third, what are the common and different points observed from each perception type? Finally, measures for reflecting residents' opinions in the process of selecting a site for locally unwanted facilities and consequent policy implications were suggested through characteristics analyses of each perception type.

The analysis was conducted through five stages: drawing statements, selecting survey targets (P sample), Q sorting, Q factor analysis, and data interpretation. The survey period was from June 1 to 24, 2015. A total of 45 out of 239 statements were drawn through literature review and expert consultation. Q sorting was conducted on 56 respondents for three weeks from July 3 to 24. Q factors extracted through analysis results were interpreted. The common and different points of each type were compared and contrasted to suggest policy implications.

Through the analysis, residents' perceptions of spent nuclear fuel repository was broadly categorized into three contexts: government trust, safety trust, and priority for environment or humans. For type categorization and naming, each researcher categorized and named the types based on the representative statement of each type. Common types were adopted through a meeting. As for points where there were disagreements, certain words that would best represent the reference that distinguished the corresponding type from the rest of the types while commonly reflecting the representative statements were suggested through discussion.

The first type (safety concerns–government distrust) was classified as having low reliability on government and safety. The second type (safety trust–government trust) was classified as having high reliability on government and safety. The third type (safety concerns–conflict avoidance) was classified as people who preferred not to have the facility in their residential areas. The fourth type (citizen participation) was classified as people who looked for citizen participation. The significance of this study lies in its identifying the differences among perception types of spent nuclear fuel repository and providing an alternative measure for policy making.

Demographic characteristics such as gender and age did not show any noteworthy difference except for the fact that the second type consisted solely of men. Conversely, a difference was observed between types concerning perception and siting of the facility. In particular, the first (safety concerns–government distrust) and second types (safety trust–government trust) demonstrated contrasting perceptions on facility safety and trust in government policy. The first type did not trust the facility safety and management and demonstrated a strong opposition to siting the repository. The second type showed a general consent to siting the repository based on their trust in government policy and facility safety. Contrarily, the third type (safety concerns–conflict avoidance) acknowledged the necessity of the facility while opposing the siting of the repository in their area of residence depending on the level of hazard. Lastly, the fourth type (citizen participation) was characterized by the residents of Gyeongju and Ulsan who viewed the facilities as necessary and stressed the importance of citizen participation in the process of decision making for site selection. These results indicated that residents' perceptions of spent nuclear fuel repository are not simply divided into agreement and opposition, but that it can rather be categorized into diverse types such as facility and government trust combined with the necessity of siting the facilities, opposition to siting the facilities in a residential area, and the importance of citizen participation. This finding is valuable as it suggests that it is possible to make decisions by taking the various types of residents' perceptions regarding the facilities into consideration rather than by a simple approval-disapproval type of local referendum when selecting the site for LULUs.

3.2. Study 4: "An Analysis of the Factors of Local Acceptance for Spent Nuclear Fuel Repository"

Spent nuclear fuel repository has almost reached its maximum capacity in Korea. The issue of safe management and conflict resolution in the decision-making process for spent nuclear fuel repository has recently drawn national attention. The purpose of this research was to examine factors determining local acceptance of spent nuclear fuel repository and to investigate moderating effects among different groups of people.

As this study aimed to investigate factors determining local acceptance of spent nuclear fuel repository and differences according to demographic characteristics, the following hypothesis groups were set to determine the relationship between measured factors and local acceptance as well as the moderating effect according to demographic characteristics.

Hypothesis group 1: The measured factors—environmental impact, economic feasibility, risk perception, sociality, and maintenance and management—will influence local acceptance.

Hypothesis group 2: There will be a moderating effect among the groups in accordance with demographic characteristics (gender, age, region, education, occupation, income, and supporting party).

The survey was conducted to analyze the residents' perceptions of spent nuclear fuel repository, and five factors were used to measure the local acceptance utilizing the structural equation model.

An online survey was conducted via a structured web questionnaire with 1,209 male and female adults across the country between August 25 and 31, 2015. In order to reduce the bias following gender, age, and region, the sample quantity was evenly assigned. The survey was repeated until the target quota was met, and a total of 1,000 answered copies were collected.

Survey questions consisted of local acceptance for spent nuclear fuel repository and opinion on the five factors (environmental impact, economic feasibility, risk perception, sociality, and maintenance and management) and were measured on a seven-point Likert scale. For survey evaluation items,

45 items were selected based on the major factors measured in previous studies. Primarily, 239 questions were drawn based on 28 pieces of existing literature including reports, books, papers, articles, and other references related to LULUs and spent nuclear fuel repository. Subsequently, 69 questions were selected based on the evaluation criteria suggested by the Korean Nuclear Society, Korean Radioactive Waste Society, and Green Korea 21 Forum (2011). Lastly, the categorized questions were incorporated or simplified through expert consultation and research meetings. Each of the items included the following contents (Kim et al., 2015): Environmental impact (five questions) included environmental issues, comfort, and sustainability; economic feasibility (seven questions) was composed of items including economic growth, income growth, job creation, population increase and social development, and economic reward; risk perception (13 questions) included factors such as human health, facility safety, and risk perception; sociality (13 questions) addressed trustability (consistency and transparency of related organizations and trustability of media and science groups) and perception on the necessity of the facilities; and, lastly, maintenance and management (seven questions) covered the imple-mentation capacity of the responsible organization and correspondence with international standards.

In order to investigate the validity of the measured factors for local acceptance, confirmatory factor analysis for each factor was conducted together with exploratory factor analysis. First, an exploratory factor analysis was conducted, which drew factors determining local acceptance based on the response rates of the 45 questions used in the questionnaire. Together with the confirmatory factor analysis for each of the drawn factors, the relationship between the factors was analyzed using the structural equation model. As a result, environmental impact and economic feasibility had a high positive relation to local acceptance rather than risk perception.

Environmental impact and economic feasibility had a positive influence on local acceptance. For example, if one perceived that the development of the facilities could be carried out without destroying the natural environment, the acceptance for spent nuclear fuel repository would be higher. Thus, it had a positive relationship with environmental impact. In the case of economic feasibility, siting the facilities revitalizes the sluggish local economy. This brings economic benefits, and therefore economic feasibility has a positive relationship with local acceptance. In terms of risk perception, even though the spent nuclear fuel repository was safe, it does pose a risk if people perceive it to be a risk. Therefore, people will perceive the siting of the facilities as negative and the local acceptance will decline. Maintenance and management is an item that evaluates the trust level in government and related managing or regulating organization. This item was contrary to the hypothesis that it will positively influence local acceptance, and no significant impact was observed at 95% confidence level.

The moderating effect was investigated by dividing people into two groups for each of the demographic characteristics in order to examine the difference in demographic characteristics when people perceived local acceptance. The moderating effect was analyzed by comparing the limited model in which there was no inter-group difference in the relationship between the four factors and local acceptance to the non-limited model. A significant difference was observed only between gender and income groups.

In particular, the effects of environmental impact were distinctively high regardless of demographic characteristics. The degree of economic feasibility for male and household income of less than 3 million won groups showed strong effects on local acceptance, and risk perception had strong effects on female and household income of more than 3 million won groups. Therefore, it will facilitate higher local acceptance in regions with a relatively more female and higher income population if sufficient understanding is properly reached concerning environmental impact and risk perception by including facts about maintenance and management rather than the economic feasibility following the siting of the facilities.

This study has academic value because it carried out systematic survey research based on the theoretical foundation suggested by existing studies by identifying the factors determining local residents' local acceptance for spent nuclear fuel repository and investigating the demographic differences.

The research is significant because it sought various factors that residents were concerned about as well as support for policy-making.

Conclusion

In this chapter, four studies introduced in this chapter measured local acceptance for low-/high level radioactive waste disposal facilities using a qualitative, quantitative and mixed methods approach.

First two studies analyzed the spatial patterns of the result of referendum for siting a low-level radioactive waste disposal facility. The facility was assigned to be placed in Gyeongju city after a competitive local referendum amongst four candidate cities in Korea in 2005. However, after the decision, many conflicts occurred between the residents living within and near the Gyeongju city. By analyzing spatial patterns of the referendum data, the first study identified that the local acceptances near and far from the facility were clustered with different values. The results of face-to-face interviews show that people near the nuclear power plant had low risk perception and the benefit of financial compensation for the districts offset the cost of potential risk. The second study analyzed the spatial distribution of the acceptance rate in each ward of the four candidate cities using several spatial statistical methods and several types of interviews. The results showed that the referendum system has a problem with spatial inequity within and across its jurisdiction.

Last two studies examined the local acceptance for a high-level radioactive waste disposal facility using qualitative and quantitative methods. As the temporary storage for spent nuclear fuel in Korea has almost reached its limit, the question of relocation for the facility has become urgent. Because perceptions on high-level radioactive waste may higher compared to low-level radioactive waste, later two studies identified perception types and factors of local acceptance for spent nuclear fuel repository. One identified people's four types of local acceptance — safety concerns-government distrust, safety trust-government trust, safety concerns-conflict avoidance, citizen participation — and differences among these perception types using Q methodology. The other study conducted a survey and identified five factors of residents' perception of spent nuclear fuel repository by analyzing structural equation model. The results showed that environmental impact and economic feasibility had a high positive relation to local acceptance rather than risk perception. In particular, the effects of environmental impact were distinctively high regardless of demographic characteristics.

These examples may provide new perspectives on management strategies of locating radioactive wastes facilities regarding local acceptance of people around these facilities.

REFERENCES

Caplan, A., Grijalva, T., & Jackson-Smith, D. (2007). Using choice question formats to determine compensable values: the case of a landfill-siting process. *Ecological Economics*, 60(4), 834-846.

Choi, J. S. (2008). Study on risk perception and acceptance for radioactive waste disposal facility after a referendum. *Korean Republic Administration Review*, 42(2), 149-168.

Chung, J. B., & Kim, H. K. (2009). Competition, economic benefits, trust, and risk perception in siting a potentially hazardous facility. *Landscape and Urban Planning*, 91(1), 8-16.

Chung, J. B., Kim, H. K., & Rho, S. K. (2008). Analysis of local acceptance of a radioactive waste disposal facility. *Risk Analysis,* 28(4), 1021-1032.

Gallagher, L., Ferreira, S., & Convery, F. (2008). Host community attitudes towards solid waste landfill infrastructure: comprehension before compensation. *Journal of Environmental Planning and Management,* 51(2), 233-257.

Getis, A., & Ord, J. K. (1992). The analysis of spatial association by use of distance statistics. *Geographical Analysis*, 24(3), 189-206.

van der Horst, D. (2007). NIMBY or not? Exploring the relevance of location and the politics of voiced opinions in renewable energy siting controversies. *Energy Policy*, 35(5), 2705-2714.

Kim, T., & Kim, H. (2010). Analysis of Spatial Patterns on the Result of Referendum for Siting a Radioactive Waste Disposal Facility in Gyeonju. *Journal of the Korean Urban Geographical Society*, 13(2), 117–128.

Kim, T.-H., & Kim, H.-K. (2014). The spatial politics of siting a radioactive waste facility in Korea: A mixed methods approach. *Applied Geography*, 47, 1–9.

Kim, T., Park, H. J., Moon, J., & Kim, T. (2015). A Study on the Perception Types and Characteristics on Spent Nuclear Fuel Repository: Focused on Q Methodology. *Journal of KSSSS*, 31, 5–25.

Kim, T.-H., Park, H.-J., & Moon, J.-W. (2016). An Analysis of the Factors of Local Acceptance for Spent Nuclear Fuel Repository. *Journal of Korea Planning Association*, 51(5), 199–213.

Kontos, T. D., Komilis, D. P., & Halvadakis, C. P. (2005). Siting MSW landfills with a spatial multiple criteria analysis methodology. Waste Management, 25(8), 818-832

Kunreuther, H., Kleindorfer, P., Knez, P. J., & Yaksick, R. (1987). A compensation mechanism for siting noxious facilities: theory and experimental design. *Journal of Environmental Economics and Management*, 14(4), 371-383.

Lidskog, R. (2005). Siting conflicts - democratic perspectives and political implications. *Journal of Risk Research*, 8(3), 187-206.

Maderthaner, R., Guttmann, G., Swaton, E., & Otway, H. J. (1978). Effect of distance upon risk perception. *Journal of Applied Psychology,* 63(3), 380-382.

Rogers, G. O. (1998). Siting potentially hazardous facilities: what factors impact perceived and acceptable risk? *Landscape and Urban Planning*, 39(4), 265-281.

Sjoberg, L. (2004). Local acceptance of a high-level nuclear waste repository. *Risk Analysis*, 24(3), 737-749. http://dx.doi.org/10.1111/j.0272-4332.2004.00472.x.

Stoffle, R. W., Traugott, M. W., Stone, J. V., McIntyre, P. D., Jensen, F. V., & Davidson, C. C. (1991). Risk perception mapping: using ethnography to define the locally affected population for a low-level radioactive waste storage facility in Michigan. *American Anthropologist,* 93(3), 611-635.

Stone, J. V. (2001). Risk perception mapping and the Fermi II nuclear power plant: toward an ethnography of social access to public participation in Great Lakes environmental management. *Environmental Science & Policy*, 4(4-5), 205-217.

Swallow, S. K., Opaluch, J. J., & Weaver, T. F. (1992). Siting noxious facilities: an approach that integrates technical, economic, and political. *Land Economics*, 68(3), 283.

Warren, C. R., Lumsden, C., O'Dowd, S., & Birnie, R. V. (2005). 'Green on Green?' Public perceptions of wind power in Scotland and Ireland. *Journal of Environmental Planning and Management*, 48(6), 853-875.

BIOGRAPHICAL SKETCH

Taehyun Kim

Research Fellow, Korea Environment Institute
#1012, Bldg. B, 370 Sicheon-daero, Sejong-si, 30147, Republic of Korea

Education: Ph.D. in Urban Planning & Engineering at Yonsei University, Republic of Korea

Research and Professional Experience:

1. Research Fellow, June 2014 – present, Korea Environment Institute, Seoul, Korea
2. Lecturer, Mar. 2013 – Aug. 2014,
   Yonsei University, Seoul, Korea
   Sungkyunkwan University, Suwon, Korea
   Soongsil University, Seoul, Korea
3. Invited Research Fellow, May 2013 – Dec. 2013, Korea Institute of Public Administration, Seoul, Korea
4. Visiting Scholar, Sep. 2011– Nov. 2012, Regional Economics and Spatial Modeling (REASM) laboratory of the University of Arizona, Tucson, USA
5. Principal Researcher, Mar. 2010 – Sep. 2011, National Institute for Disaster Prevention, Seoul, Korea

Publications from the Last 3 Years:

1. Kim, T., & Kim, T. (2017). Smart and resilient urban disaster debris cleanup using network analysis. Spatial Information Research, 1–10. https://doi.org/10.1007/s41324-017-0088-4
2. Kim, T., Sohn, D.-W., & Choo, S. (2017). An analysis of the relationship between pedestrian traffic volumes and built environment around metro stations in Seoul. KSCE Journal of Civil Engineering, 21(4), 1–10. https://doi.org/10.1007/s12205-016-0915-5
3. Seung-Woo Son, Jeong-Ho Yoon, Hyung-Jin Jeon*, Jong-Min Lee, J.-W. M., & Kim, T.-H. (2017). Environmental Information System Analysis and Evaluation of Information Applications for Supporting the Environmental Inspection Practices. Journal of the Korea Academia-Industrial Cooperation Society, 18(2), 726–736.
4. Kim, T.-H., Park, H.-J., & Moon, J.-W. (2016). An Analysis of the Factors of Local Acceptance for Spent Nuclear Fuel Repository. Journal of Korea Planning Association, 51(5), 199–213.
5. Jung, S., & Kim, T. (2016). Capacity Assessment of Landfill Site by Input-Output Analysis: Focused on Sudokwon Landfill Site. Journal of Korean Official Statistics, 21(2), 95–117.
6. Kim, Taehyun, Hyun Joo Park, Jiwon Moon, and Taehyun Kim. (2015). A Study on the Perception Types and Characteristics on Spent Nuclear Fuel Repository: Focused on Q Methodology. *Journal of KSSSS*, 31, 5–25
7. Kim, T. (2015). Applications of Carbon Footprint in Urban Planning and Geography. In S. S. Muthu (Ed.), The Carbon Footprint Handbook (pp. 163–182). Boca Raton: CRC Press.
8. Kim, Tae-Hyun. (2015) Linking and Utilizing Urban, Environmental, Disaster Prevention Spatial Data for a Climate Change Adaptation Spatial Planning, *Journal of Environmental Policy*, 14(1): 85-112
9. Shin, Jin Dong, Mi Sun Kim, Tae Hyun Kim, and Hyun Joo Kim. (2014) A Conceptual Review of Resilience from a Disaster Perspective - Focused on the Joseon Dynasty -, *Korean Review of Crisis & Emergency Management(KRCEM)*, 10 (12):93-107
10. Sung Jin Noh, Tae Hyun Kim, Hong Kyu Kim. (2014) An Analysis of Recognition on Application Elements of Local Identity in Urban Public Design - Focused on Pedestrian Roads in Kwangjin-gu and Songpa-gu, Seoul -, *Seoul Studies*, 15(1): 35-49

11. Taehyun Kim, Hongkyu Kim. (2014) The spatial politics of siting a radioactive waste facility in Korea: A mixed methods approach, *Applied Geography,* 47: 1-9.
12. Taehyun Kim, Sandy Dall'erba (2014) Spatio-temporal association of fossil fuel CO_2 emissions from crop production across US counties, *Agriculture, Ecosystems and Environment*, 183: 69-77.

INDEX

A

accidents, 72, 153

accuracy, 23, 25, 26, 32, 34, 92, 94, 107

activity, viii, 5, 42, 44, 66, 67, 71, 72, 78, 79, 81, 83, 84, 86, 87, 88, 89, 90, 91, 92, 93, 102, 103, 104, 106, 107, 108, 109, 110, 111, 112, 117, 143, 144, 149

activity concentration, ix, 66, 79, 81

activity estimation, 87, 88, 89, 106, 108

activity per week, 108, 109, 110, 111

angular efficiency, 100

animals, ix, 127, 128, 129, 133, 135, 136, 138, 141, 144

as low as reasonably achievable (ALARA), 5, 8, 112

attenuation coefficient, 33, 100

authority, viii, 49, 66, 68, 72, 154, 161, 164, 165, 166

B

biological hazard, 85, 86

box source geometry, 104

BRICS countries, vi, vii, x, 147, 148, 150, 151, 153, 172, 175

C

calculations, ix, 16, 32, 35, 66, 86, 88, 89, 92, 95, 99, 103, 108, 109, 141

calibration, viii, 2, 10, 24, 25, 26, 27, 38, 53, 98, 112

categories, 6, 10, 27, 45, 68, 85, 163

cellular stress levels, 135

characterization, 74, 79, 134

classes, 76

classification, 67, 71, 73, 74, 76, 77, 78, 79, 80, 81, 82, 114

Index

clearance, viii, 65, 76, 78, 79, 80, 81, 82, 83, 84, 86, 88, 108, 109, 110, 111, 113, 114, 115, 118

clearance level, viii, 65, 78, 79, 80, 82, 83, 84, 86, 88, 108, 109, 111, 114

clinical source, 108, 110

concentration, 35, 68, 71, 73, 80, 81, 85

concentration limits, 80, 85

contamination, 72, 83, 111

control, 4, 29, 33, 34, 38, 52, 72, 73, 74, 77, 79, 80, 82, 111, 115, 116, 118, 121, 149, 155, 156, 159, 160, 163, 164, 169, 172

controlled areas, 72

correction factor, 30, 98, 99, 100, 102, 113, 133, 137

corrections, 15, 89, 90, 99, 101, 103, 108, 109, 113

criteria, 6, 18, 48, 49, 55, 57, 67, 69, 72, 77, 78, 81, 154, 163, 192, 196

D

decay, 81, 84

decay storage, 81, 84

decontamination, 72, 154, 170

detection, 87, 99, 103, 104, 111, 117, 118, 140

detector, ix, 3, 34, 54, 66, 86, 87, 88, 89, 90, 98, 101, 102, 103, 104, 105, 106, 107, 109, 110

developed countries, 47, 111

discharge, viii, 66, 73, 79

disposal, vi, vii, viii, x, 65, 68, 72, 73, 74, 78, 79, 81, 85, 86, 88, 111, 114, 116, 117, 147, 148, 149, 151, 152, 153, 155, 156, 157, 162, 163, 165, 166, 167, 168, 169, 170, 171, 172, 174, 175, 177, 178, 181, 195

disposal system, 85

distances, 106, 110

distinct geometries, 104

---

dose, vii, viii, 1, 3, 4, 5, 7, 8, 9, 10, 12, 13, 14, 18, 19, 20, 21, 22, 23, 24, 27, 28, 29, 30, 31, 32, 33, 34, 35, 36, 37, 38, 39, 40, 41, 42, 44, 46, 47, 48, 49, 50, 51, 52, 53, 54, 57, 58, 61, 62, 63, 65, 78, 82, 83, 86, 87, 88, 89, 90, 91, 92, 104, 117, 118, 134, 141

dose calibrator, 89, 90, 91, 92, 117

dosimetry, vii, viii, ix, 2, 13, 17, 18, 23, 24, 27, 28, 29, 30, 31, 32, 33, 34, 36, 37, 38, 39, 40, 41, 53, 56, 57, 59, 112, 122, 123, 124, 127, 128, 138, 141, 142, 143, 144

E

effective doses, 42, 78, 82

effective energy, 99, 106, 107

electron, 12, 94, 95, 96, 97, 98, 101, 102, 112, 125

emission, 67, 88, 92, 100, 104, 148, 149

empirical formula, 95

energy, ix, x, 10, 11, 12, 13, 20, 23, 26, 30, 32, 36, 37, 39, 41, 49, 54, 55, 56, 57, 63, 66, 83, 86, 92, 93, 94, 95, 96, 97, 98, 99, 100, 101, 102, 103, 104, 105, 106, 107, 112, 113, 114, 115, 116, 117, 118, 119, 124, 125, 128, 138, 141, 147, 148, 149, 150, 153, 154, 155, 156, 157, 159, 160, 161, 163, 164, 165, 166, 168, 169, 170, 173, 174, 175, 177, 195

energy absorbed by the animals, 138

energy dependency, 99, 100, 103

energy spectra, ix, 66, 95

ensure safety, 69, 80

environment, viii, x, 2, 10, 27, 30, 50, 53, 66, 67, 68, 72, 73, 74, 77, 111, 112, 114, 116, 120, 148, 149, 157, 162, 164, 170, 172, 179, 188, 189, 192, 197, 198, 199

environmental law, 148

Index

experimental, vii, ix, 13, 86, 89, 90, 93, 99, 100, 104, 120, 124, 127, 128, 129, 130, 131, 132, 133, 134, 135, 136, 137, 138, 141, 143, 144, 159, 196
exponential decay, 90
exposed persons, 67, 72
exposure rate, ix, 66, 86, 87, 88, 89, 90, 91, 92, 93, 99, 100, 103, 104, 106, 107, 109, 111, 112, 119, 125
exposure/dose, 86

F

facilities, vi, vii, viii, x, xii, 2, 9, 10, 24, 65, 69, 74, 77, 79, 87, 88, 92, 104, 125, 153, 156, 157, 158, 160, 161, 162, 163, 164, 167, 168, 169, 170, 179, 180, 181, 185, 186, 188, 189, 190, 192, 193, 195, 196, 197
finite differences in temporal domain (FDTD), ix, 127, 128, 129, 130, 133, 136, 137, 138, 141
fluctuation, 92, 107

G

gamma, 58, 67, 72, 78, 83, 88, 97, 99, 100, 103, 104, 112, 118, 119
Geant4, 103, 104, 124, 125
geologic repository, 148, 174
geometrical, 50, 87, 88, 89, 92, 99, 101, 106
geometry, ix, 27, 34, 66, 88, 89, 90, 100, 104, 105, 106, 107, 108, 109, 110, 111, 112, 121, 125
Getis-Ord General G, 187
Getis-Ord Gi*, 184
GTEM, ix, 128, 130, 131, 132, 133, 136, 137, 141
GTEM chamber, 132, 133, 136, 137, 141

H

half-life, viii, x, 65, 67, 72, 78, 81, 91, 109, 148
health, 72
health physics, 72
hellium gas, 102
hight sensitivity, 101
histological changes in the rat thymus, 136
human, 24, 66, 74, 192
human health, 24, 66, 74, 192

I

IAEA regulations, 69, 71, 80
incident, 10, 14, 15, 16, 21, 27, 28, 30, 31, 38, 39, 97, 98, 100, 131, 132, 137, 164
influences, 20, 106, 107, 111, 182
instrumentation, 111
instruments, viii, 2, 10, 14, 24, 26, 38, 57, 84, 86, 87, 118
inverse square-law, 87
ion pair, 94, 95, 96, 97, 98, 101, 102, 112, 119
ionization, 13, 23, 24, 27, 30, 34, 36, 45, 57, 98, 104, 113, 114
ionizing radiation, vii, ix, 1, 2, 3, 4, 5, 6, 7, 8, 22, 27, 28, 45, 46, 47, 53, 54, 55, 59, 67, 113, 128, 140
ions, 96
irradiation, 27, 106, 107, 108, 118, 131

K

kinetic energy, 11, 96, 97
Korea, xi, 151, 179, 180, 181, 186, 188, 189, 191, 192, 194, 196, 197, 198, 199

L

laws, x, 148, 150, 152, 153, 161, 170

legislation, vii, x, 83, 148, 150, 153, 154, 157, 159, 163, 164, 170, 172, 173, 174, 177

limit, xi, 4, 67, 71, 73, 80, 93, 94, 99, 104, 180, 194

limitations, 32, 42, 51, 52, 78, 81, 87, 89, 92, 101, 102, 140, 141

local acceptance, vii, x, xi, xii, 179, 180, 184, 185, 188, 191, 192, 193, 194, 195, 196

locally unwanted land uses (LULUs), x, 179, 181, 186, 190, 192

M

management, vii, viii, xi, xii, 2, 3, 6, 8, 48, 49, 51, 52, 53, 62, 65, 66, 69, 73, 74, 77, 79, 80, 84, 85, 86, 87, 88, 104, 111, 113, 115, 116, 117, 118, 123, 149, 151, 152, 155, 156, 157, 163, 165, 166, 167, 169, 170, 171, 174, 176, 179, 180, 181, 190, 191, 193, 195, 196, 197, 198

manufacturer, 90, 99, 101

mass, 99, 108

mass attenuation coefficient, 99, 108

mean energy, 12, 94, 95, 96, 97, 98, 101, 102

mean SAR values, 137

measurement, ix, 14, 15, 16, 23, 24, 27, 30, 31, 35, 36, 38, 62, 66, 83, 86, 87, 89, 90, 100, 102, 103, 109, 111, 113, 117, 131, 133, 138, 141, 144, 184

medical, vii, 1, 2, 3, 4, 5, 6, 7, 8, 27, 53, 55, 73

medical areas, 73

medical exposure, vii, 1, 2, 3, 4, 5, 6, 7, 8, 27, 53, 55

medicine, viii, 65, 85

minimize animal stress, 129

monitoring, 34, 49, 52, 83, 115, 131, 145, 162

Monte Carlo, ix, 32, 35, 37, 58, 66, 89, 92, 103, 104, 120, 122, 124, 125

Monte Carlo simulation, ix, 33, 66, 89, 103

Müller, Geiger, ix, 66, 86, 89

multifrequency, ix, 128, 131

multifrequency System, 131

N

national authorities, 24, 74, 77

neon gas, 101, 102, 104

non-point-source, 89, 106

nuclear industry, 155, 166, 172

nuclear law, 148

nuclear medicine, viii, 2, 3, 6, 7, 8, 9, 28, 29, 41, 42, 61, 65, 66, 85, 112, 117, 120, 125

nuclear medicine facilities, viii, 65, 85

O

occupational workers, 77

occurrence of electric discharge in the brain, 134

optimized, 4, 7, 9, 80

overestimated, 109

P

package, 78, 83, 85, 86, 88, 89, 109

permissible levels, 67, 72

photons, 10, 20, 32, 93, 96, 97, 99, 104, 105, 108

physical half-lives, 111

plane waves, 133

point-source, 86, 87, 88, 89, 91, 92, 100, 102, 103, 104, 105, 106, 109, 110

Index

practice, viii, 5, 7, 8, 9, 10, 12, 21, 25, 26, 27, 28, 30, 31, 34, 40, 50, 53, 56, 57, 66, 68, 78, 80, 82, 86, 87, 104, 111, 120, 156

probability, 19, 20, 47, 78, 82, 95, 99, 104, 105, 106, 108

protection, vii, 1, 2, 3, 4, 5, 6, 7, 8, 9, 22, 28, 53, 54, 55, 56, 57, 58, 59, 60, 61, 62, 67, 69, 74, 76, 79, 81, 83, 113, 114, 115, 116, 118, 120, 121, 124, 149, 154, 155, 161, 166, 170

publication, 4, 5, 6, 34, 47, 48, 54, 55, 57, 59, 60, 61, 62, 69, 81, 98, 124, 150, 153

Q

Q methodology, xi, 180, 189, 194

quality control, 90, 92, 94, 123

R

radiation, vii, ix, x, 1, 2, 3, 4, 5, 6, 7, 8, 9, 10, 11, 13, 14, 15, 16, 20, 21, 22, 24, 25, 26, 27, 28, 29, 30, 32, 33, 34, 35, 36, 37, 38, 39, 40, 41, 42, 45, 46, 47, 48, 49, 50, 51, 52, 53, 54, 55, 56, 57, 58, 59, 61, 62, 63, 67, 69, 72, 74, 76, 78, 80, 82, 83, 85, 86, 87, 89, 97, 99, 101, 104, 112, 113, 114, 115, 116, 117, 118, 119, 120, 121, 123, 124, 125, 127, 128, 129, 130, 132, 133, 134, 136, 137, 138, 141, 142, 143, 144, 148, 154, 161, 162, 163

Radiation, 83, 86

radiation dose, vii, 1, 2, 3, 8, 21, 33, 35, 37, 38, 39, 40, 41, 47, 48, 49, 62, 63

radiation level, 67, 72, 78, 82, 83, 85, 87, 142

radiation monitoring, 83, 86

radiation protection, vii, 1, 2, 4, 5, 7, 10, 20, 24, 26, 34, 48, 53, 54, 55, 57, 78, 86, 113, 162, 163

radioactive decay, 79, 81

radioactive materials, 29, 66, 68, 80, 152, 155

radioactive waste, vii, viii, x, xi, xii, 65, 66, 67, 68, 69, 72, 73, 74, 76, 77, 78, 79, 80, 81, 82, 84, 85, 86, 87, 88, 89, 104, 111, 112, 116, 117, 125, 147, 148, 149, 150, 152, 153, 155, 156, 157, 158, 162, 165, 166, 167, 169, 170, 171, 172, 173, 174, 175, 177, 179, 180, 181, 188, 194, 195, 196, 199

radioactive waste disposal, viii, x, 66, 69, 77, 148, 150, 155, 166, 169, 170, 177, 179, 181, 194, 195

radioactive waste disposal facilities (RWDFs), x, 179, 181, 186, 194

radioactive waste facilities (RWFs), vii, x, xii, 179, 180

radioactivity concentration, 83

radioisotope, 68, 87, 91, 92, 93, 99, 104, 106, 107, 116, 117

radiological, 4, 5, 6, 7, 8, 9, 10, 19, 27, 29, 47, 49, 54, 55, 57, 58, 59, 60, 61, 62, 67, 68, 77, 81, 82, 112, 154

radiological commissions, 82

radiological protection, 4, 8, 19, 48, 55, 69, 77

radionuclide, viii, x, 29, 65, 78, 81, 85, 86, 88, 89, 92, 113, 148

radionuclide concentrations, 78

recommendation, 4, 18, 20, 68, 71, 72, 73, 74, 77

referendum, xi, 179, 181, 182, 184, 185, 186, 188, 190, 194, 195

reflected power, 131, 132

regulation, 67, 68, 71, 73, 76, 77, 113, 155, 157, 161, 163, 172

regulatory body, 77, 80, 82, 154, 155, 164

regulatory control, 74, 78, 80, 81, 85

release, 67, 72, 73, 80, 82, 84, 86, 108, 109, 110, 111, 116

releasing, 72

removal, 80, 82

reports, 45, 68, 87, 174, 192
residual, viii, 65, 84, 85, 86, 87, 88, 112
residual activity, viii, 65, 84, 85, 86, 87
responsibilities, 77, 152, 154, 157, 164, 172
risk, viii, 2, 22, 29, 53
risk assessment, viii, 2, 22, 29, 53

S

safety, xi, 3, 4, 8, 9, 55, 56, 66, 67, 68, 74, 76, 77, 80, 81, 109, 111, 114, 115, 116, 117, 152, 153, 154, 155, 161, 162, 163, 164, 168, 170, 174, 176, 180, 186, 188, 189, 190, 192, 194
safety management, 66, 80
SAR, ix, 128, 129, 133, 134, 135, 136, 137, 138, 141, 142, 144
SAR for simultaneous radiation with 900 and 2450 MHz signals, 137
scattering, 106, 122
scenario, 87
segregation, viii, 65, 73, 81
seizures, 134, 143
sensitive volume, 92, 98, 99, 101, 102, 103, 105
short lived, 76, 78, 81
simulation, 90, 92, 104, 105, 109, 123, 124, 125, 128, 129, 133, 134, 137, 139, 144
solid, vii, viii, 11, 65, 66, 67, 72, 73, 74, 81, 82, 83, 84, 85, 87, 88, 89, 92, 100, 102, 103, 112, 114, 118, 121, 125, 195
solid angle, 11, 92, 100
solid radioactive wastes, vii, viii, 65, 66
solid waste, viii, 65, 67, 69, 73, 74, 77, 80, 82, 83, 85, 87, 88, 89, 102, 111, 118, 152, 155, 157, 164, 172, 177, 195
source, 3, 7, 8, 15, 41, 52, 89, 90, 92, 95, 97, 100, 103, 104, 105, 106, 107, 108, 109, 110, 118, 174
spatial analysis, 180, 185, 186, 188

specific absorption rate (SAR), ix, 128, 129, 133, 134, 135, 136, 137, 138, 141, 142, 144
specify limits, 73
spectra, 92, 95, 99, 103, 104
spectrum, 12, 16, 25, 49, 50, 88, 99, 104, 129, 130, 131, 132, 133
spent nuclear fuel repository, xi, 180, 188, 189, 190, 191, 192, 193, 194
staff, 86, 120
standard sources, 89, 92
standards, 10, 24, 25, 28, 57, 66, 68, 76, 80, 81, 94, 95, 114, 115, 116, 118, 119, 154, 163, 170, 192
standing wave cavity, ix, 128, 134, 135, 138
storage, vii, viii, ix, xi, 65, 66, 67, 72, 73, 77, 78, 81, 83, 84, 85, 86, 88, 89, 103, 108, 109, 111, 116, 117, 149, 153, 157, 162, 163, 167, 168, 169, 170, 180, 194, 196
storage period, 85
storage time, ix, 66, 85, 86, 89, 108, 109, 111
stress, 135
surface, 13, 14, 15, 16, 18, 21, 28, 30, 31, 32, 33, 36, 39, 52, 58, 79, 83, 86, 88, 105, 106, 107, 108, 109, 110, 111, 162
survey, xi, 49, 101, 113, 180, 184, 189, 191, 193, 194

T

technical report, 56, 73, 152
technique limitations, 103
theoretical deterministic formulae, 87
travelling wave cavity, 128
treatment, 83
treatment requirements, 83

Index

U

uncontrolled disposal, 79

V

validation, 104, 122, 124, 125, 138, 141
volume direction, 101

W

waste, x, 78, 80, 83, 85, 86, 87, 88, 108, 111, 162, 179, 181, 186, 194
waste disposal, x, 179, 181, 186, 194
waste generation, 74
waste management, 67, 69, 73, 74, 77, 80, 85, 88, 111, 118, 152, 155, 157, 164, 172, 177
waste management, 80
waste management, 162
waste management system, 80, 162
waste package, 78, 83, 85, 86, 87, 88, 108, 111